D0927747

BETTER BEE KEEPING

BETTER BEE KEEPING

THE ULTIMATE GUIDE TO KEEPING STRONGER COLONIES AND HEALTHIER, MORE PRODUCTIVE BEES

KIM FLOTTUM

QUARRY BOOKS

BEVERLY MASSACHUSETTS

First published in the United States of America in 2011 by
Quarry Books, a member of
Quayside Publishing Group
100 Cummings Center
Suite 406-L
Beverly, Massachusetts 01915-6101
Telephone: (978) 282-9590
Fax: (978) 283-2742
www.quarrybooks.com
Visit www.Craftside.Typepad.com for a behind-the-scenes peek at our crafty world!

ISBN-13: 978-1-59253-652-8
ISBN-10: 1-59253-652-2

Digital edition published in 2011
eISBN-13: 978-1-61058-028-1

Library of Congress Cataloging-in-Publication Data available

Design: Burge Agency
Illustration: page 156, Peter Sieling

Printed in China

Dedication

I've always thought of a dedication in the front of a book as an acknowledgment and a thank you to those people who, by unselfish acts, assisted the author in some way ... as a mentor, teacher, supporter, or faithful friend.

But, there are people who do those things because they are supposed to.

H. G. Wells once said, "There is no human urge greater than the urge to change someone else's copy." It turns out though, that even great writers need editors. And the rest of us need really great editors. They see the wheat in the chaff and the trees in the forest.

I've been lucky. First there was Winnie Prentiss, who simply asked if I'd like to write a book about beginning beekeeping. They wanted one, and could I do it? She helped with *The Backyard Beekeeper*.

Then along came Rochelle Bourgault: A tiny sprite of a thing, who writes, teaches yoga, and knits in her nonediting time. She has shepherded *The Backyard Beekeeper's Honey Handbook*, the revision of *The Backyard Beekeeper*, and now this work, too, from beginning to end. She even went to a beeyard with me and stopped to smell the honey and the bees.

If I'm lucky, she'll punch up this small effort. An editor wouldn't want just an average piece of work telling readers that the book is good because she helped. It should be as good, and maybe better, than the rest. Thanks, Rochelle.

Contents

Introduction

Dr. Hachiro Shimanuki, Research Leader, USDA Honey Bee Research Laboratory, Beltsville, Maryland, Retired

Some years ago I attended a lecture given by Dr. Hachiro Shimanuki, who was then the research leader of the USDA Honey Bee Research Laboratory in Beltsville, Maryland. Shim, as he is called, was discussing current problems beekeepers were having and the research his lab and others were involved in looking for answers to those problems.

I don't recall the particulars but when he was nearly finished he offered an observation that was, for the most part, missed by the majority of those present. They were more interested in his answers. Answers and dogma, went the feeling, saves bees, money, and time. Observations, on the other hand, were believed to be for spectators and scientists without a vested interest in the day-to-day workings or the bottom line of a beekeeping business. Those attending didn't want an either/or answer, but rather a simple yes or no.

Shim's observations were, however, profound, and any beekeeper who listened carefully to his challenge is probably doing quite well today.

Basically, his observation was this: He called it the Rule of Rights. It is the inspiration and basis of this book.

If you produce the right number of bees that are the right age and in the right condition, and are in the right place at the right time, you will be successful.

The complexity of achieving this goal is well hidden in the simplicity of his statement. But to accomplish this requires making intelligent and correct decisions based on sound planning, correct timing, and getting the balance of business and biology to work in an operation. There's little how-to hidden within this simple statement. Rather, it is a goal to strive for in many ways. It is, in the real world, not easy and it is not often that it is achieved. But when it is achieved, it is a thing of beauty.

What follows on these pages is my perspective of the application of Shim's rule, applied to areas of beekeeping that are generally the most difficult to do well, and thus often overlooked. Primarily, we'll apply this rule of rights to producing honey, producing queens, overwintering in cold areas, and population control during the active season, which means swarming. And of course, there is the requisite section on dealing with the pests we all face. But these, too, can be viewed through the lens of the right rules.

The construction of a bee beard as large as this one is a perfect example of the rule of rights: the right number of bees (30 pounds, or 13.6 kg); of the right age (collected from the brood nests of many colonies so they are young and strongly attracted to a queen); in the right place (on Dr. Norman Gary's body); in the right condition (the bees have been held queenless for several hours, so are strongly attracted to the vials of queen pheromone around his neck); at the right time (had the bees been required to remain queenless for longer, or shorter, the attraction would not have been as strong).

The information offered here comes not from any wisdom I have gained as a beekeeper for the thirty-plus years I've been lighting smokers and opening hives, but from the hundreds, no, the thousands of beekeepers I have had the good fortune to know, observe, and learn from. They, not I, provide the wisdom and experience you will find here. Rather, my role is to gather, analyze, evaluate, and offer the information gleaned from those who worked to produce it, and then only after it has passed through some set of filters manned by those who still offer answers and make observations. Wisdom, it seems, has many sources. In the end you can pick and choose just the right bits for you, at your pace, sized and priced for your operation.

No matter the scope of your operation, whether 5, 500, or 50,000 colonies, getting the most from your bees requires a delicate balance of business skills, well-managed labor, a fundamental understanding of the principles of honey bee biology and health, a keen sense of place, and cooperative weather, all mixed together with perfect timing and a little bit of luck.

There is another rule of rights that should be added to this list: that what needs to be done with the bees is done the right way at the right time. Tom Theobald runs two-queen colonies, and if he isn't on top of supering at the right time he'll miss a lot of his honey crop. In an area with finicky crops, missing even a little can be the difference between success and failure for the season.

What Do You Need to Know Before You Begin?

To benefit the most from the information here you need to be comfortable with several concepts and practices:

- You should be comfortable with both the science and the art of making queens and be familiar with basic colony and honey bee biology and nutrition.
- You should understand principles of nectar and pollen manipulation, and be able to deal effectively and, more important, safely with pests.
- You should be producing both varietal and artisan honeys, followed by late-season management to prepare for the down season (if there is one), particularly cold and confined winters.
- You should understand marketing and plan for business expansion.

To be certain these concepts are clear and to ensure that you recall the fundamentals of all of these topics we provide background and basics throughout these pages as easy reminders that reinforce these important biological, management, and business concepts. Look for them for additional and supplemental information.

There is a lot in this book that you may already know, think you've heard before, or have at least suspected for some time. There's comfort in that: Much of what we know, or are pretty sure we know, as beekeepers we've figured out on our own because we mostly work on our own. Seldom are there "experts" at our side when we are confronted with a problem in the field. And though the beekeepers you are in touch with routinely may be good. I'd predict that these particular sources are just as often as unsure of the answers as you are.

You may find that you are a much better salesperson than a honey producer or a much better soap and hand cream maker than beekeeper. But you won't know until you try your hand at them all, and explore other sides of having honey bees in your life.

So like the beekeepers at that meeting so many years ago, I'm quite sure you are here for answers, rather than observations. You will not be disappointed. There are some answers here. They are straightforward, practical, and intuitive because the old adage about change is true: Not much has changed in the past thirty years. Still, know that this is not your everyday how-to book. There are many of those already.

But those pesky observations are here, too. They tend to be hidden in plain sight, because though not much has changed, how we accommodate those challenges and situations is changing.

So I challenge you to grow by being certain of the answers you already know, and to grow stronger by learning from the observations of the thousands that have shared what they know.

And all the while, keep your veil tight, your hive tool handy, and your smoker lit … because, don't you know, next season will be better.

Chapter 1:

Growing Your Operation:

How Big Do You Want to Get?

Monoculture

There was a popular song by a rock 'n' roll group called the Outsiders in the 1960s (or so I'm told) with a title that perfectly describes what most businesses endure as they grow. The song, *Time Won't Let Me,* was about romance, but the lyrics can easily be applied to trying to keep ahead of all those bees you have now. What started out as a passionate affair with a box full of bugs can, and often does, turn in to a never-ending, always-growing task. And if not resolved, it will quickly cause more trouble than an impatient lover.

I can't wait forever
Even though you want me to
I can't wait forever
To know if you'll be true
Time won't let me ... no
Time won't let me ...

From *Time Won't Let Me,* written by Tom King and Chet Kelley, *Capitol Records,* 1966

When your hobby turns into part-time work, but the "parts" keep getting bigger, the push and pull between your beekeeping time and the rest of your life increases. Making a conscious decision to change what it is that you have now to something else will add another dimension of stress to the situation. Without enough preparation before you begin and careful attention during the transition, you can muck up both the bees and the rest of everything you do: your family, your day job, the things you do for fun, and potentially, the future of your career with the bees. Since you've invested this far in this work, I'm comfortable with the assumption that the business part of your bees is producing enough money, and maybe even profit, to take it seriously, and that your goal is to further expand the business in some direction.

Garden of Eden

Trading Time for Money

You know the saying about planning to fail and failing to plan. I doubt you are doing the former and I hope you're not doing the latter. Here are some thoughts on planning and goals when dealing with bees.

You may have guessed that I'm basically lazy. But it takes money to be lazy. You should consider cultivating a specific kind of laziness, too. The underlying message is that you probably do not put a fair value on your time.

Buy Preassembled Frames

Assembled frames with plastic foundation are more expensive than unassembled frames with beeswax foundation that you put together. And even though I apply wax to that plastic foundation, it still takes me less time to prepare those frames than assembling frames and inserting and wiring the foundation. But to me, the time I save is more valuable: That is, I make more money during the time I save not assembling frames than I would have saved if I had purchased regular frames and spent that time assembling them.

I purchase assembled frames that have plastic foundation already inserted. You have to apply some beeswax to the foundation, which takes time for preparation and application, but it makes them more appealing to the bees. And it takes far less time than wiring in beeswax foundation. Plus, waxed plastic foundation is just fine for the bees when it's done correctly. I spend more money for these frames, but I save a lot of time.

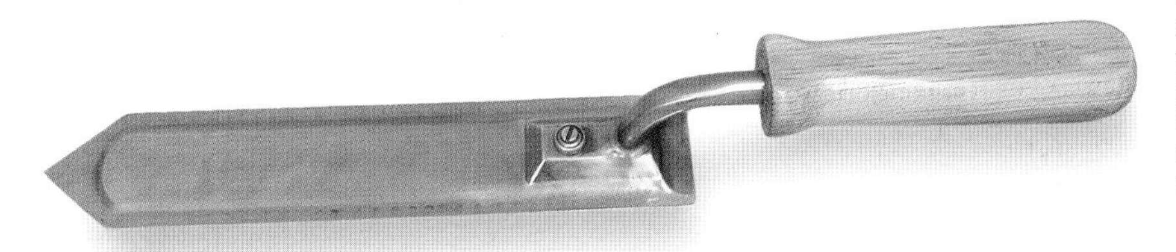

Uncapping Knife (cold)
After using a cold knife for uncapping his first year, my friend tried a hot knife. It wasn't any better. So he traded money for time and paid for an expensive uncapper, but he saved enough time that he actually made more money.

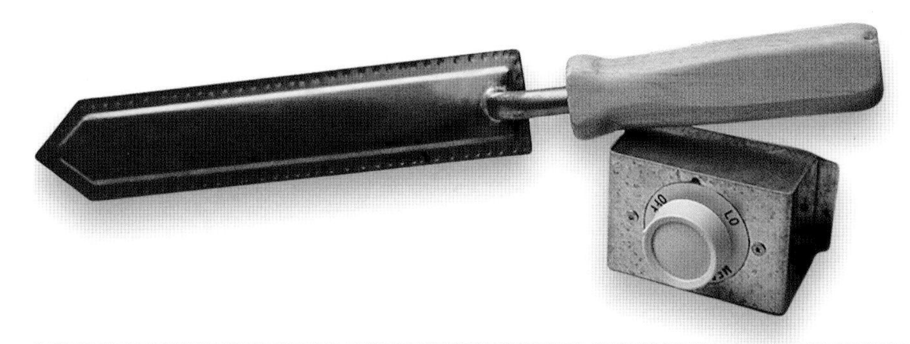

Uncapping Knife (hot)
It turns out that a hot knife wasn't any better than the cold knife for saving time and work.

Planning Ahead Saves Time and Money (But You Knew That)

Another example of using your time wisely (and "lazily"): A friend of mine started out with a couple of hives and enjoyed it so much he got more full-size hives the first year, and even more hives the second year. At first he had help from another beekeeper when harvesting and processing. He uncapped his harvest the first year by hand using a cold knife. That first year he knew neither what to expect nor what his options were. The second season he was far more successful, and had more hives in the beeyard and had many more frames to handle. He'd done some homework and thought that using a hot knife would be faster than using a cold knife. But after a few hours with that hot knife he saw the light. He pledged, "Never again."

"Uncapping by hand," he said, "takes too much time away from my other tasks."

But he didn't quit keeping bees, nor did he go out and get the first automatic something he could find just to make life easier in the short run.

Instead, he spent a lot of valuable time studying the art, science, and mechanics of harvesting, processing, and handling honey. He talked to other beekeepers, manufacturers, and suppliers. He read books, went to meetings, and figured out what size he wanted his operation to eventually reach. He looked at all the equipment and space he was going to need—a loading dock, hot room, box moving equipment, an uncapper, and an extractor that could handle the right number of frames. He studied not just a wax melter but also a wax processing system. He checked on the plumbing for a water heating and moving system, and looked at lighting for night work and windows and skylights for day work.

Setting up a honey extraction facility is not work for the faint of heart or the unskilled. Do not underestimate the value of a trained plumber, electrician, and builder when you undertake this task.

He outlined the sumps, pumps, pipes, strainers, heaters, coolers, and bottling machines he'd need to bottle honey. He studied the code, requirements, and wiring for all the electricity he'd need. Plus, he looked at the tanks he'd need to store honey, and figured out storage for everything from boxes to bottles to tools to his truck and even an office and conference room—all of it.

In other words, to plan a better uncapper, he essentially built his future honey house in his head, based on his goals for growth. And then he heeded the advice of nearly everyone he talked to and doubled the space his original plan allotted.

And though his was still a small-scale backyard operation, that's the uncapper (next page) he bought in year three. Way, way more than his eighteen colonies needed, for sure. (Plus, he had to borrow money to do it.)

His philosophy is similar to mine. In his mind the interest paid on that loan was equity he was putting back into the business because every year that uncapper gave him more time to spend with his bees, with his business, and with his family. After a few years and other equipment purchases, he even had time to do custom extracting for other beekeepers, which led to making money with the time he saved.

Needless to say, he never picked up an uncapping knife again, and after fourteen years he's exactly where he planned to be and darned if that uncapper isn't just the perfect fit. That's how I look at growth in this business: It starts with the right plan.

Use It Up, Wear It Out

Richard Taylor was a noted philosopher, commercial beekeeper, author of several best-selling beekeeping books, and a regular columnist for beekeeping magazines for decades. As a child he lived through the Great Depression and in spite of his frugal youth went on to an exceptional academic life. As a result of his background, education, and beekeeping experiences, Richard professed a different philosophy when it came to time, money, and his beekeeping business: "Use it up, wear it out, make it do, or do without." This mantra was given to him by a strict mother in lean times. In his opinion this philosophy was a major factor in his beekeeping success. This concept is not original to Richard, but it reflects a more traditional view of trading time for money: The more time you have to invest in your business the less money you will have to spend to be successful. If time is not a luxury (and many jobs' schedules allow this), trading in much more time and spending much less money is a perfectly good plan.

Knowing the value of the time you have is a gift. Use it wisely.

Up to now I suspect you have been trading time for money to a great degree. You must know the value of your time: what you could be doing instead of building frames, uncapping by hand, or lifting every super four or five times. If what you would rather be doing is watching TV, then read no further because you don't need what's offered here. But if using that time to raise better queens, sell your honey at an exclusive farm market, negotiate a lease to plant a couple hundred acres of clover, or make up nucs from colonies running over with bees seems like a better investment in your business and your future, then read on.

TIP:
It's not always possible to make my philosophy work, but I tend to look at situations requiring cash (such as an uncapper) and my mind first asks, "Rather than cut expenses somewhere, how can I increase income somewhere?" Can I raise prices, sell unused equipment, produce another product to sell? Can I do something smarter or faster or charge more for it so I have the same amount of available time, but generate more income in the process?

When envisioning the uncapping process, you need to consider how your system will work when there is one person running all the equipment, and then when there's more than one person. It takes some planning, but in the long run a smooth-running honey house saves an incredible amount of time. And time is in shortest supply during honey season.

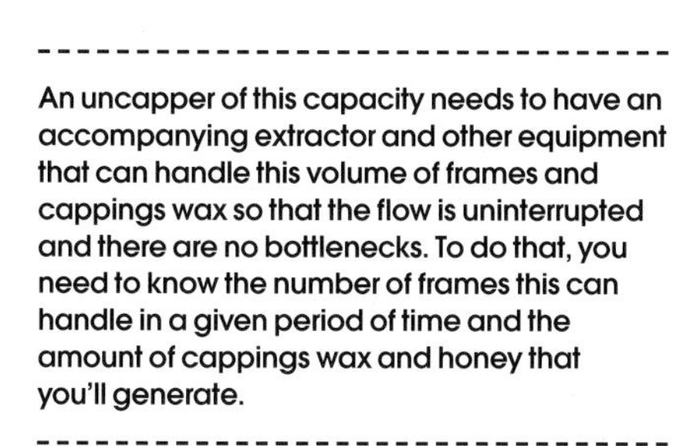

An uncapper of this capacity needs to have an accompanying extractor and other equipment that can handle this volume of frames and cappings wax so that the flow is uninterrupted and there are no bottlenecks. To do that, you need to know the number of frames this can handle in a given period of time and the amount of cappings wax and honey that you'll generate.

This beekeeper took a stainless steel milk tank and converted it into a honey holding tank. He had to add,a spigot, piping, and other connections, and it had to fit in his honey house. He was able to purchase it for much less than a typical honey tank, and was able to do the work himself, saving a significant amount of money in the process. However, this wouldn't work for many beekeepers because of the space required, and the skills needed to safely install it. This is a good solution to a local problem, but with limited adaptability to other operations.

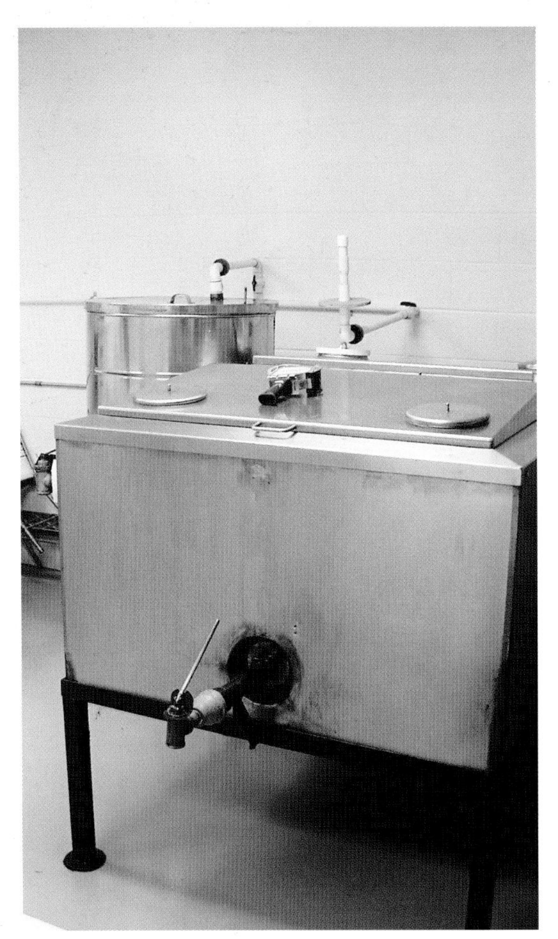

This handy cart was made from a two-wheeler with a couple of extensions, some welding, and better tires. When used to load beehives onto a low trailer, it does the work of a second person who would otherwise be doing half the lifting. This beekeeper spent time making the cart, but saves time every time he had to move bees, and saved money by not having to hire help. This is a good solution to a problem that is adaptable to a more generalized beekeeping crowd.

I run a tight ship financially. It's far from perfect, but my payback from spending more to accomplish something, to meet a goal, almost always eventually pays for itself by saving me time in both the short run and the long run. Time is—without question—what I need the most of, have the least of, and want more of.

An exception to this time/money conundrum can be found in the tool kits of creative and talented beekeepers. A friend's son once commented, "All manner of gadgets can be made in a beekeeper's workshop to enhance almost every aspect of beekeeping. They don't need to work extremely well, nor be well made—yet the bees seldom mind, it gives the maker a sense of accomplishment, eases some of the burdens of the craft, and often saves the maker some money." He's right, you know—as long as you consider the value of the time it takes to make the gadget.

The truly creative are those who spend time rather than money to make improvements to make things that save time, energy, and money again and again. That time is well spent.

Standardized Equipment, Genius Ideas, and Patents

In the thirty-some years I've been in this business I've seen countless inventions that solve big and little problems in beeyards, in honey houses, and even moving bees—the list goes on and on. Many widgets, as ingenious as they are, only solve a problem (or fix a mistake) unique to a particular situation. Often, transferring the concept or technology to other locations is neither as efficient nor as time-saving and seldom works. The common thread for something that is creative is that the concept and form—whether it's a painting, poem, computer program, or tool—should be original, new, and useful.

Occasionally, there are ideas that transcend specific situations and can be applied to other beekeeping locations, situations, or experiences. Here are some opinionated thoughts on this process.

■ **First**, even if you conceive of a widget or process that absolutely every beekeeper should have, and your invention is specific to solving a beekeeping problem, there simply aren't enough beekeepers to make you rich. So the question becomes: How useful is it?

■ **Second**, when you conceive of, and then make, that widget, and even prove to yourself and others that it will solve a host of beekeeping's woes, the thought of a patent often comes up—too often. It is relatively easy to do a patent search on the Internet, and spending money on an idea that won't make you much money is a practice to seriously consider avoiding. Most likely you'll find something very similar that has been around for years. So ask yourself these questions: Why isn't anybody making, using, or selling it now? Is my idea not new, not useful, or not original? Can I really be the first beekeeper to have this problem and solve it this way?

■ **Third**, even if you don't find something similar, spending money on it is seldom a good idea:
a) It most likely was tried years ago and was discarded because of a flaw in the material, the design, the concept, or the technology adaptation.
b) There is probably already a cheaper, faster, better way to solve the same problem your widget solves, so look first.
c) If it is indeed new, and can be easily or inexpensively copied, beekeepers, like most farmers, are generally inventive enough that they'll make their own once they see how you did it and will probably do it better.

■ **Fourth** is a common sense check: If you have enough money to find and prosecute everyone who copies your low-cost gadget, then you don't really need the money you thought you were going to make selling these in the first place.

Avoid the temptation to get rich quick by producing yet another beekeeping widget. Stick to what you know, what you can do, and what you like. Leave inventions to the inventors.

However, I am not advising you not to be as creative as you need to be, save as much time as you can by doing so, and share your truly new idea with the world. You can always share your idea in a letter to the editor in one of the beekeeping journals. The good you'll do there will always outweigh the money you thought you could make.

Business Goals and Bigger Pictures

One day, you will reach the point of asking yourself, "Is this still a hobby?" Or, you will reach a goal of observing, "I'm big enough now to take the next step." What is the next step? What's your plan?

If you haven't already, you must develop an informal business plan. A formal business plan or business model may be necessary later, but your banker will let you know all about that. My informal business plan starts with three simple questions: Where are you now? Where do you want to be? How will you get there?

Successful beekeepers never sit and let things happen. Successful beekeepers go out and happen to things.

1. Where Are You Now?

This is really a list of what you have—a detailed summary of who you are in a material way. Mostly. Basically, what do you have? List all your beekeeping "stuff," and by each item list an approximate value. Some things are easy to put a price on, such as a used truck. Some things aren't, like a beeyard location. But try anyway.

Consider the range of your beekeeping "stuff": Vehicles. Farming and gardening implements. Tools, lumber, and everything stored in the garage or shed. Buildings and land. Beeyard arrangements. Sales outlets.

Also consider your range of financial responsibilities. Savings. All debts—credit cards, mortgage(s), college tuition (this year and the next), truck loans. Don't forget that future debt you have planned, too.

When considering the things that you have, count everything you have at home or the shop, and don't forget what you have in all your beeyards—colonies, fences, batteries, and more. Is there honey in those boxes? That's future income. Count that too.

Reality check: When you wager the future of your business and your family on the weather and the bees, exercise some restraint, get good advice, and talk to as many beekeepers as possible to best manage your expectations.

You probably own more than one smoker. If you're going to expand, you will definitely own more than one. If you are using a smoker every day, a smoker will last about a year, whether it's made of expensive stainless or cheap galvanized (unless you run it over with your truck). Why buy stainless? Because that's all the manufacturers make.

Consider income: current income from all sources (including from the bees). Future income (like royalties from a book you are writing). Include your family's income.

The list goes on and on. Write it all down, with no priorities and no holding back or leaving off or forgetting. Set it aside and wait a day or a week. Let your family look at it, or even your attorney and banker. If you have a few good friends who understand what you do, show your list to them too. Once you've listed all of your material assets and expenses, add it all up. Accountants call this your net worth, your assets minus your liabilities. Your total may be a negative number. That's okay because you need to know.

An objective financial analysis goes a long way in helping you decide your future. If the ink is red and changing it by using means other than the bees seems remote, your choices are pretty straightforward, but by no means are they cast in stone.

Even if the ink is solid black, the vagaries of the weather can spoil even the best of plans. As a general rule, financial planners suggest that when you are starting a business, assume you'll have no profit for three years or so. (You will have income.)

Your next calculation is for what the income you are generating covers: all, some, or none of your current responsibilities (other than your beekeeping business)? If it is only some or none, how will those outside expenses be covered?

Disclaimer: I'm not a financial planner. Reality check: When you wager the future of your business and your family on the weather and the bees, exercise some restraint, get good advice, and talk to as many beekeepers as possible to best manage your expectations.

Amidst all this material tallying, do not lose sight of the issue of time and money. When you begin this growing process, if you invest more time than you are currently able to commit, the income from the period before you turn a profit will be shortened, probably significantly. A friend at work who is a financial consultant once said that three years is sound advice, but don't accept it as gospel without a fight. By increasing your cash flow, reducing your spending, and perhaps relying on outside family income, you can shorten that three years to months, or to no time at all. Providing, of course, the weather is good.

I Don't Like Doing This ...

One aspect of the "Where Are You Now?" question is making sure you list your likes and dislikes of what you are doing and want to do. There's probably not a financial cost to these items, though there might be, but the grief and the joy they add to the equation are worth noting.

If you're keeping bees just so you can run your supply store, make soap and candles, help the local bee club, and sell honey and everything else to the public, maybe keeping bees is taking time you could be spending more profitably. Conversely, if waiting on customers who complain about your prices and keeping your accountant happy keeps you up at night, maybe you should be spending all your time with the bees. Whatever it is, if you hate doing it, it is costing you stress and grief. Don't do it.

2. Where Do You Want to Be?

Where do you want to be, and when do you want to get there?

Most people start by answering with something simple, just to get their head around the concept, such as "Bigger." "Independent." "Rich," maybe. Then they fine-tune it "in two years, or five." "Profitable enough to quit my day job." "Enough bees to make sure I can keep all my customers all year long." "To sell enough queens to pay for a new honey house, or a truck, or my kid's college education." Twenty-five more hives, that's all, just twenty-five.

Making a ton of honey may be your goal. But if the discipline of keeping bees drives you nuts, and processing and packing honey is so much fun that it keeps you making honey, maybe that's all you should be doing. Or if selling everything you can make and buying more to resell is what gets you up in the morning, but working bees all day makes you crazy, listen to that voice. Maybe you hate selling, dealing with people, handling cash, asking for money, but being with the bees is the best thing you've done since you got married. If so, there's a solution for you.

One very good way to answer this is to ask those closest to you: "What do I complain about the most?" More often than you'd suspect they'll be right on the money and a little more than eager to let you know. And if you're honest with them, and yourself, you'll listen and learn. But deep down, you already knew.

So back to the question, "Where do you want to be?"

Whether it's making soap, creams and lotions, or candles, when using products from your bees is more interesting, more fun, and more profitable than simply keeping bees, you have an easy choice to make. Bees shouldn't be the excuse you use to do what you really want to do, and what you're really good at.

A Funny Thing Happened on the Way to My Next Career

Procrastination interferes with accomplishing most tasks, but sometimes it works to an advantage.

While going through the "Where do I want to be?" exercise, a friend kept changing and refining and honing his list. At the time he was a moderately serious backyard beekeeper and he wanted to get a lot bigger, but the investment in time and money wasn't working as well as he wanted. The lack of money required to fund the adventure kept making him refine those goals. All along, his one-way-to-get-more-from-his-bees trick was using some of his honey and wax to make soap, and that just slowly kept expanding. Soon he started making labels for other soap makers because his were so good. Then he started selling his soap wholesale to distributors. Next he started making and selling his soap with someone else's label. Next he was selling soap-making supplies to other soap makers. And then he was selling other products to his many soap outlets … and not long after he was in the Everything Soap business, making a small fortune and having a wonderful time doing it. But he still had only a few colonies and was still trying to figure out how to be a bigger beekeeper.

When the lightbulb finally went on, he realized the value of his soap business, and rediscovered the joys of backyard beekeeping.

One Thing to Do: Feed the Foodies

The interest in urban beekeeping has helped fuel the demand for "locavore" food—food sourced from somewhere nearby.

All restaurants use honey, and high-end eateries can use several hundred pounds a year in their special desserts and sauces. Having two or three colonies on the roof or out back is a marketing tool some restaurants are already using, and more will do so when they see the business it brings in and the money it can generate. I suspect that the money they save on the honey produced on site will hardly equal the cost of keeping those colonies there, but it's an investment they will continue to make for the cachet.

You can try to make a part-time living from this in several ways:

- They can pay you to manage hives they purchase, and they get all the honey;
- They can let you keep *your* bees on *their* property and buy your honey from your hives;
- Some combination or other arrangement. The permutations are nearly endless and they all are profitable for the manager of the hives and the users of the honey.

When making the pitch for rooftop hives, convincing the owner is one thing (and too often the owner is simply a business consortium of doctors or lawyers that doesn't want anything out of the ordinary), and convincing the chef is another. My vote goes for the chef—and sharing recipes that come from a chef of equal caliber to the one you are talking to can only help.

(For three chef's quality recipes, see pages 24–25.)

City or Suburban Sharecropper

Keeping bees in the city has taken on a certain glamour, but let's face it, keeping bees in the city isn't for everyone.

There are many challenges to keeping bees in the city, and on rooftops of places you don't live. Parking, for starters, can be an issue, along with just getting stuff to where you need it. Gaining easy access to the roof and removing honey supers full of honey down a hallway (maybe leaving a few bees around hallway light fixtures) can be an interesting challenge. The constant wind on rooftops can hinder flight, and colonies will need good protection in the winter, and easy access to water will test the mettle of any colony. Roofing materials may generate a fair amount of heat. On the other hand, cities generally have lots of water, so abundant forage from early to late season and good honey crops are common. Interestingly, a colony's isolation from other colonies in a city will reduce the incidence of pest and disease problems.

You, however, can be the beekeeper for all manner of city dwellers—government agencies wanting to be green, owners of rooftop gardens, managers of city garden spaces, libraries, museums, people with more money than common sense who simply want to be able to say "Yes, we have bees in the garden (or roof, out back, on the deck)."

Each of these is a potential income source. But know the liabilities of moving bees in and out and the potential hazards of harvesting, vandalism, and safety.

There's a whole different set of issues when bees are kept at street level. Vandalism and public safety are always issues. But more and more urban areas are changing their way of thinking and are allowing colonies to coexist with their urban neighbors. Those who keep bees in the city, or allow you to help them, are contributing to the general biodiversity of the urban landscape, helping to pollinate the many street trees and plants and gardens, and of course are able to harvest "their" honey. You may even discover there's more money to be made keeping bees in town than in the country.

Handled correctly, there is money and honey to be made as an urban sharecropper.

There's a whole different set of issues when bees are kept at street level.

Three Restaurant Recipes from Michael Young, MBE

So your beekeeper guests have arrived for dinner, and you know you need to impress them. Well, you know they are sick of mead (no doubt they only get the early stuff without any vintage). The trouble with mead is that it's so damn good to drink most people don't let it mature enough. Right then, let's make a cocktail fit for a king and a beekeeper—one that will tickle the pants off them.

Honeysuckle Cocktail

For the recipe:

5 teaspoons (25 ml) freshly squeezed lemon juice

1 teaspoon (5 ml) mead

1 teaspoon (7 g) sourwood or similar honey

7 teaspoons (35 ml) Black Bush Irish whiskey

4 teaspoons (15 ml) fig liqueur

1 pound (455 g) ice block

Small lemon wedge

Method:

1. Squeeze the lemon juice 1 hour in advance of serving.
2. Mix the mead and honey together.
3. Add the whiskey, lemon juice, fig liqueur, and honey–mead mixture to a shaker.
4. Add the whole block of ice to the shaker, and shake well.
5. Pour through cocktail strainer.
6. Pour into a frozen 3-ounce (90 ml) cocktail glass and garnish with the small lemon wedge.

Yield:

1 cocktail

Honeyed Pumpkin Velouté with Honey-Toasted Pumpkin Seeds

For the recipe:

7 ounces (200 g) pumpkin flesh

1/2 cup (50 g) chopped onion

1/2 cup (50 g) chopped leek, white part only

1/2 cup (50 g) chopped celery

3 tablespoons (42 g) butter

1 tablespoon (20 g) honey

1 quart (946 ml) white vegetable stock

1 teaspoon (5 ml) lemon juice

Good-quality salt and pepper to taste

8 teaspoons (40 ml) double cream

1 egg yolk

9 pumpkin seeds

Michael Young, MBE, teaches culinary skills at the Belfast Metropolitan College in Belfast, Northern Ireland. A master chef for more than forty years, he has won many gold medals for his culinary skills, which include food sculptures and dishes using his own honey and mead. He is also a senior British honey judge, as well as a Welsh, Scottish, Irish, and Georgia (U.S.) judge. He is a master beekeeper in the University of Georgia's Beekeeping Program. For his efforts and dedication to the advancement of apiculture worldwide, Michael was awarded the MBE, the Order of the British Empire, in 2008.

Method:

1. Peel, seed (pumpkin flesh is called for) and chop the pumpkin.
2. Sweat and caramelize the vegetables in the butter and honey with the chopped pumpkin, in a heavy-based saucepan.
3. Add the stock. Simmer until all the ingredients are fully cooked.
4. Purée and pass through a fine strainer.
5. Add the lemon juice. Adjust the seasoning and consistency.
6. Whisk the cream and egg yolk in a bowl until mixed.
7. Add half of the hot soup to the yolk and cream, and then add the liaison back to the soup. (Be sure to add only half the soup; if you add the soup all at once you will scramble the eggs.)
8. Bring to just under a boil (but do not boil or the soup will split).

Garnish:

Add a knob of butter and 1/2 teaspoon (3.5 g) of honey to a hot pan. Add the pumpkin seeds, moving them constantly until the seeds are toasted. Transfer to a dry paper towel. Add as garnish to the soup just before serving, with a very small drizzle of honey on top of the soup.

Yield:

4 servings

Marinate-Infused Sea Bass with Wok-Fried Oriental Vegetables and Aromatic Honey-Flavored Sauce

For the sea bass:

Four 5-ounce (142 g) pieces sea bass

For the marinade:

- 2 tablespoons (28 ml) olive oil
- 2 tablespoons (40 g) strong honey
- 3 tablespoons (45 ml) light soy sauce
- 2 tablespoons (28 ml) hoisin sauce
- 2 teaspoons (12 g) rock salt
- 1 star anise
- 6 cloves
- Juice of 1/2 lemon

Method:

1. Place the fish in a small deep bowl or tray.
2. Mix the marinade ingredients together, and then pour over the sea bass.
3. Leave in a cool place for 30 minutes.

For the vegetables:

- 2 teaspoons (10 ml) sesame oil
- 2 teaspoons (10 ml) walnut or nut oil
- 2 teaspoons (14 g) honey
- 1 carrot, cut into thin strips
- 1 red onion, cut into thin strips
- 4 tablespoons (50 g) grated ginger
- 2 stalks lemongrass, white part only, crushed and finely diced
- 5 cups (250 g) beansprouts
- 1 pok chow, cut into rough pieces

Method:

1. In a wok over high heat, combine the sesame and walnut oils and honey. Gently place the marinated sea bass into the hot oil for a few minutes. Turn and sauté until cooked through.
2. Gently remove the sea bass from the heat, place on a paper towel, and keep warm.
3. Bring the wok back up to a hot temperature.
4. Carefully add the carrot, onion, ginger, and lemongrass. Cook for a few minutes, making sure the wok stays hot and constantly turning the vegetables.
5. Add the beansprouts and cook for about 30 seconds, then add the pok chow, stirring loosely for 1 minute. The whole cooking operation of the vegetables should not take more than about 4 minutes; you want the vegetables to remain crunchy.

To serve:

1. Transfer the cooked vegetables to a hot plate.
2. Gently place the sea bass on top.
3. Return the wok to the heat to reheat the sauce, then drizzle over the sea bass and a little around the dish.

Yield:

4 servings

Raising queens is a time-honored way to supplement your beekeeping income. You can raise them for yourself (see chapter 3) plus you can produce queen cells to sell by themselves, to requeen your colonies, to stock your nucs, or to stock the nucs you sell. Raising good queens will save you time and money in your operation, and make you money when you sell them

What Can You Do?

Your research and previous experience will tell you what kind of income you can expect to earn with the skills you have or are willing to learn from the 50, 500, or even 5,000 colonies you may soon have.

- Income can be from all directions, such as simply producing and then selling just your honey, or packing both your crop and the honey from other beekeepers and selling under your label. You could buy and pack varietal honeys from various locations to broaden your line.
- Crop pollination is another way to make money, using your own trucks or renting, leasing, or hiring transportation.
- With the high-profile attention that good nutrition has, collecting pollen to use or sell should also be considered.
- Rendering and selling wax, or making something from beeswax (yours or from others), remains a source of off-season income.

Once you have enough colonies, raising most of your own queens becomes possible. The natural extension of this is selling extra queens or queen cells. Making and selling early-, mid- and late-season nucs is an option when you have enough colonies. Plus, you can stock those nucs with your queens or queen cells, then put them in boxes you make off-season, or in ready-made boxes you have made locally, or in the customer's own box. For increased sales you can offer your nucs with someone else's queens or queen cells, maybe trading your queens for theirs.

The package industry is still in business and as a commodity they will always exist, but as a business model it is declining under its own mismanagement.

Making and selling nucs, in my opinion, is the future of the beekeeping industry.

You can make all or most of your own equipment if you have the tools and workshop space, time, and skill. If making your own is possible you can even sell a few special pieces you are particularly good at or have the material for.

An often-overlooked source of income to consider, especially when investing time instead of money in your operation, is custom honey extraction for other beekeepers. This can be a moneymaker, especially from beginning beekeepers who have yet to invest in the equipment. Calculating costs is straightforward: The time it takes (do you pick up and deliver?) from beginning to end has a cost in actual dollars. But there can be additional fees for atypical jobs: handling lots of bees left in the honey supers, sorting through lots of uncapped frames, uncooperative (broken, Superglued-in) frames—the list is endless. I call these irritant fees, but you might have a better name. Consider and calculate wear and tear on your equipment too.

Then, you can determine an agreed-upon payment in honey, money, assistance in some form for your operation, or some other cash or barter agreement. Ethics and honesty are important, though: Without them you'll be out of business in a season.

You can charge per box, per box size, or per hour. Experience helps determine which route is best for you.

What about leasing 100 or 1,000 acres of land that you will work yourself, or pay to have it worked for a honey crop or beeyard? How do you spend the winter season? Do you winter where you are, keeping your bees inside, or outside, but pack your hives? Do you buy or make the feed you use for some or all of your colonies, both sugar and protein, every season or only in emergencies? Like queens, can you make more than you need and sell to other beekeepers? A mixer and equipment that's used for ten times what you need will pay for itself a lot faster than if it sits unused most of the year.

After this kind of self-evaluation you may discover, like our soap-maker friend, that you are not headed toward the bees. You can let go of the idea that you do this because you love bees, and focus (embrace is not a word I would ever, ever use … well, here anyway) on what you really love. If having bees is simply your excuse to get to do those other, more enjoyable jobs, you will save yourself and probably a lot of other people a lot of grief.

When I suggest you ask others what you complain about, and ask you to ask yourself what it is you want to do, be honest. Then figure out how to do what you do best and enjoy the most, because that's where you want to be.

Now, you know where you are and where you want to be. What comes next is more difficult.

3. How Will You Get There?

To get to where you want to be requires both time and money, with the goal of spending the least amount of money necessary to make the most amount of money possible. And achieving this goal should take up no more time than you're spending with the bees now. Time costs money, and when you are spending more, it should be offset by an increase in income. For the sake of this book I'm assuming (with all the dangers that may bring) that your expansion is toward bees and beekeeping, unlike our candle and bee supply store friends whom we met earlier.

If it's only honey you want to make, set up your operation to handle barrels. The income and labor costs per pound (0.45 kg) will be the lowest you will experience, but you'll still have queens, pollen, wax, even propolis to harvest with all that honey. Plus, there's all those nucs to sell.

Making Money by Removing Bees

The demand for people who can remove honey bees from places they shouldn't be has always been with us, but now the demand is growing. Colonies of wild bees are slowly increasing as natural resistance to varroa grows in the feral bee population, and all those new beekeepers haven't figured out how to stop swarms. At the same time, many homeowners are having second thoughts about simply killing honey bees and are trying to save them instead.

A common scenario is when a homeowner calls a beekeeper with carpentry skills and asks about the bees that are in his wall.

If you have the necessary skills, removing bees from locations where they are unwanted can be a lucrative addition to your income, with the potential to be a major source of income, plus a never-ending source of bees—bees that are clearly healthy enough to thrive without a beekeeper's help. It is a specialized skill requiring both knowledge of bee behavior, and experience with building and rebuilding walls and other structures. You'll need business acumen, an investment in equipment most beekeepers will never use, enough time, and good people skills.

Bee Removal Requirements

Specialized tools, plenty of time, good insurance and attention to detail are what you need if you are going into the bee removal business. But the rewards can be profitable, and the work is mostly seasonal: very busy in the spring, even more so in the summer, when nests begin to get larger and there are more bees to be noticed, and very slow in the winter. In many situations there is a lot of honey and wax to harvest from within the walls. If the job is done correctly, you almost always get the bees, plus you get paid.

Most jobs require half a day, seldom more than a day. To remove bees this efficiently you really need all the right tools and often someone to help. Quotes are made over the phone by most removal professionals, and advertising is almost completely word of mouth. Of course, some jobs are too big, or too complicated, and you may just have to walk away. So before you begin, get some experience, and know when to say no.

Licensing and Permits

Discussions of licensing, permits, and pesticide use come up frequently with this job. Removing stinging insects, by killing them or physically removing them, usually precipitates this question. Pest removal regulations vary by state and county, as do home repair laws. Check them all out. You may be required to take and pass an exam to qualify to be a certified pest remover, whether you ever use chemicals or not. If you decide to make this a substantial part of your occupation, don't quibble: Take the test, pay the fee, get the insurance, buy the tools, and decide what you'll do with all those free bees.

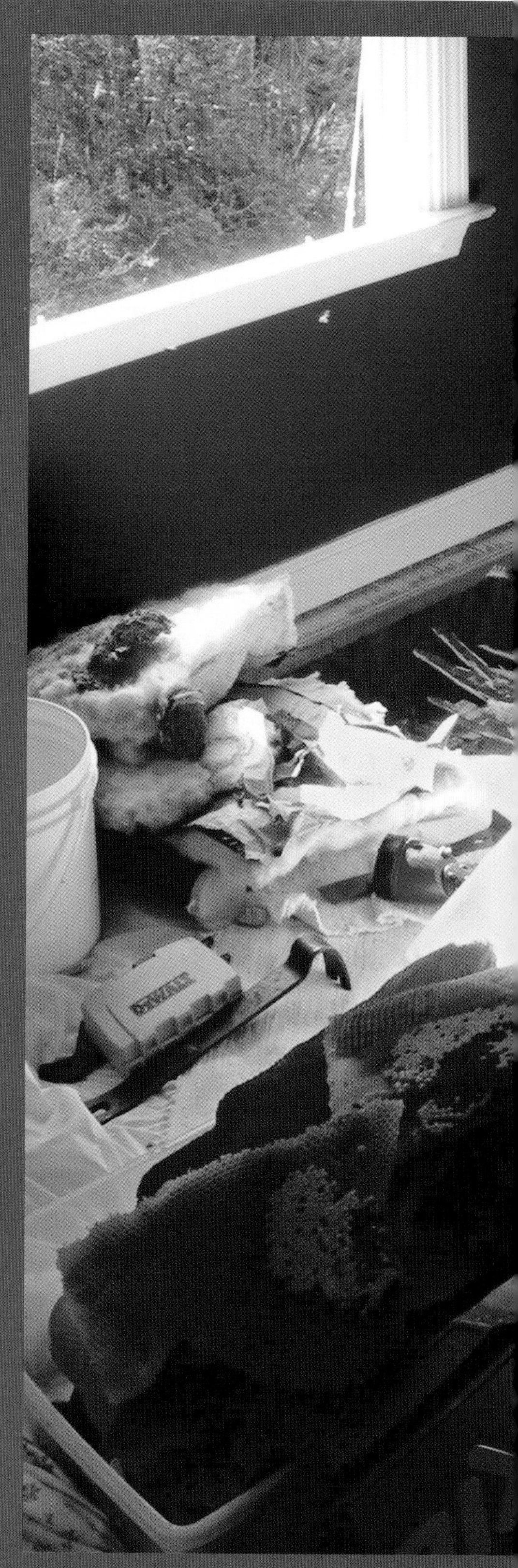

Special tools, plenty of time, good insurance, and attention to detail are necessary if you are going into the bee removal business.

Unbuilding a House

I recommend one excellent book on the subject of basic and not-so basic types of building construction, which also details techniques of dismantling buildings and building parts: roofs, soffits, walls, attics, floors, basements, columns, windows, and even trees. And then it shows how to put them all back together again.

Removing Bees, by Cindy Bee and Bill Owens, published by the A. I. Root Company, teaches you about the equipment you'll need (to buy or build yourself), and discusses the business considerations of a bee removal occupation, from quotes to liability to customers who don't pay.

If this side of honey bees appeals to you, absorb all available information, talk to other removers, start slow, and read the book.

Pollination

You may have been approached by a grower to move bees to his farm to pollinate some of his crops, or by another beekeeper to supplement his colonies to fulfill his contract to pollinate a crop. Maybe it worked and you made good money with little effort. Maybe you just broke even but you had cash flow at a slow time of the year. Or maybe it went horribly wrong.

Pollination responsibilities require:

- Colonies to be at a minimum population of adults.
- Open and sealed brood to be present.
- Colonies to be in good health.
- Availability on a certain date.

Nothing New There

However, managing colonies to make all these factors happen on schedule is the test of a master beekeeper. If you've been involved you know the challenge. If not, as you grow and additional income streams become necessary (and possible) that challenge will be tempting.

Managing bees for pollination is different than managing bees for producing a honey crop—mostly. Trying to do both with the same bees is often a cause of honey bee and beekeeper distress. Elsewhere it is mentioned that this manner of management required an entirely different skill set and instruction manual. That's still true.

From the business side, having a good pollination contract is necessary. And not having one is an invitation to a suicide. Beekeepers who have been in the business for years will tell you that a contract is not necessary and more trouble than it's worth. They report that some growers won't use one and that it's a flat-out statement that you don't trust the grower you want to do business with.

When you are being given this advice, listen carefully, nod your head, mumble something about they're probably right, and run. My advice is to not take that advice.

A contract will protect you, your family, your business, your employees, and your future. Ignore it at your peril.

Contracts are discussed in the chapter on leasing land to produce a honey crop. They are even more important here. Examples of pollination contracts abound on the Internet. Look at all of them. Include everything that's important to you, the grower or broker, and most important, to your business.

Cover Your Assets

"Cover your assets"—CYA—is all-important when someone else has a say in what happens to your bees. What is the liability for theft, for stinging incidents, either with employees or with non-employees? Most important, how will you recover if it all goes bust? Incidences of theft, someone switching your good equipment for someone else's junk, abandoning your equipment, or selling equipment that belongs to you occur every year to beekeepers unwilling to listen to this advice. And mostly it's not reported because the beekeeper who got cheated or robbed is too embarrassed to speak up.

To Move or Not to Move ...

It's true that many beekeepers manage their bees for both pollination and honey production. They do this by moving from warm wintering areas to honey-producing summer areas and pollinating crops in between. My perspective is that this amount of activity asks too much from a colony of bees. Some beekeepers ask this from their bees every season.

Severe queen and colony losses are common for multi-move migratory operations. Any move causes some percent queen loss, and making many moves in a year is too stressful for a colony.

If you're going to winter in one place, make that place a pollination job without an additional move. Or, make your summer destination a pollination job, then settle in for the honey crop.

Most commercial beekeeping operators consider my opinion wrong. Too cautious, even cowardly, they say, and besides, it's the only way to make a living with the bees. They're your bees. Anything you decide is fine.

Moving bees for pollination is hard on your bees and your equipment, and, if you are moving them by hand, hard on your help. If you undertake this task, be aware of the hidden costs involved: getting your colonies strong enough, early enough, with extra feeding and labor; moving costs; queen loss during moves; and the opportunity cost of lost honey production both during the pollination contract and during the recovery afterward. If you build all of these costs into the rental price, plus some extra for unforeseen incidents, plus the profit you need, then pollination is a good deal.

Some other questions that should be answered in a contract:

- Are you protected if the grower dies and your bees are sitting on his land?
- What if the corporation that owns the cropland is sold, even to a foreign investor or a housing developer?
- What if a neighboring farmer sprays his crop and your bees are in that crop? And what about damage from pesticides, whether the grower's or someone else's?
- How is your price arrived at? Do you know your full costs? There are costs involved with keeping a colony strong. There are significant transportation costs to consider, too. Do you settle for the price that is offered or negotiate?
- What is the payment schedule?
- How much notification will you be given to move in and out?
- Are there provisions for acts of God, such as that 100-year storm?
- What happens if your colonies are stolen from the orchard?

A final word on contracts: If a grower, broker, or other beekeeper is unable or unwilling to negotiate with you to ensure all of the above concerns are protected, here's what you'll end up doing: Explaining to your banker why your payment is late; getting your family to understand why your business is bust; and telling your employees that unemployment is the future because you trusted somebody with a handshake. Have the other party sign on the bottom line, or leave. Period.

Renting to Rent

Recently, ambitious beekeepers have hit on an interesting way of handling pollination rentals. They rent large quantities of bees from much larger, migratory operations. A commercial interest can provide you with bees that meet the specs you require and drop them off at your holding yard. You, then, parcel them out to growers near you, handling the business side of this as if they were your own. The contract is the same. In turn, you have a contract with the beekeeper that arranges strength, delivery date, pickup date, payment, and so on.

The advantages of this system for you are awesome. The margins are small, but the cost is negligible. None of the typical colony management costs apply. All you do is distribute them, transport them, collect the fees, and pay the commercial operation. If you carefully estimate your delivery, inspection, and pickup costs, allowing room for anything else that can go wrong, you'll know what you can afford to rent them for, and how much to charge.

From the commercial operator's perspective, the advantages are also profitable. The rental rate to you will be higher than to a grower because of the smaller scale of the job. But the cost will be essentially the same as to a grower, so the profit margin increases. The contract may spell out certain care and liability issues that you will have to agree to, but that's simply being safe on both sides.

Contingencies for weather delays must be considered, and a Plan B, should the commercial operator have a disastrous winter, must be kept in mind. This may add stress and cost to the deal. Overall, it is a business model to consider for helping cash flow in the spring and to establish more customers for your pollination business as your operation grows.

This beekeeper found a used storage tank at an auction that was too big for most operations, but it fit his new building perfectly. Plumbing was simple, the condition of the tank was exceptional, and he was able to get it on the cement pad inside before the walls went up—perfect timing for a perfect piece of equipment.

Buying Equipment

Reducing overall costs is an option, but buying anything cheap can be a gamble. Purchasing used equipment may be the exception because it can save you a lot of money. You may have made used equipment purchases but mostly from people you know. If you are handy at repairing things, then spending time repairing, and maybe re-repairing older equipment may be the best way to go. Again, it's time, or money. Is the time you spend fixing something costing you less than the cost of a new or not-as-used item would have been?

Answer that question when examining used equipment for sale. And don't make rash decisions just as the proverbial auctioneer's gavel is coming down. Buyer beware.

An operation that is expanding is a good place to look for good equipment. That beekeeper did enough things right to be able to grow, and most likely he has been taking care of his equipment on a daily basis.

There are also equipment brokers, classified ads in the journals and on the Internet, and the ever-present beekeeper's underground that can feed you information on available equipment and the condition it's in. Listen to as many beekeepers as you can find for leads and look for the occasional going-out-of-business auction, but make your own decisions on quality.

Buying out backyard beekeepers is a good way to get basic equipment, boxes primarily. You can look at frames and spot irksome foulbrood right away, so you will know if it's worth checking further or not even worth considering. The most important aspect of buying older boxes is the question of time versus money. You'll spend less money on used boxes, but you may spend more time fixing their problems. For example, if used boxes come with frame spacers, and you don't use them, they're a nuisance and you have to spend valuable time removing them.

Standardization and Value

The two things you already know are essential: standardizing the size of boxes you use and the value of the boxes. Small-scale operators tend to overprice their equipment, so have a good feel for what you can pay for used stuff. Don't overpay. But keep buying boxes because without them you don't have any place to put all that honey.

One metric some beekeepers use is that you should be able to use at least 75 percent of the boxes you buy the day you buy them. The rest are fix or burn. Frames, however, are evaluated on a different metric. If you get 25 percent you can use, and feel good about using them (they are clean, not broken, less than jet black, the right size), you are in good shape. If you assume that if you buy 100 boxes, you'll get to use 75, and only 25 of the boxes will have frames, you can set your paying price there. It's conservative, but it's safe.

The Mercy Buy

Everybody knows it exists. Nobody admits it. A beekeeping operation folds for whatever reason and the owner is in dire straits financially. When the stuff comes up for sale, you are presented with a mercy buy. For purchases like this, figure that you will get half of what you would normally—and remember that someday it might be you on the auction block.

Places to Put More Bees

When more boxes are in your future, more places to put them will be, too. Finding a good beeyard is a process that takes at least a year, and two is better. Elsewhere in this book there is an in-depth look at developing your own yards through leasing resource-rich land. That is the best way to solve the beeyard problem. But, again, it's a question of time or money. If leasing isn't in the cards (the cost of leasing may be the issue as well as available land), then new beeyards are in your future.

There's one aspect of finding beeyards that isn't often discussed. If fact, never in my experience has a beekeeper actually suggested that it's a good idea to pay good money for a good beeyard. The old axiom of "we take the beeyards we can get" is a sad truth. And a dumb one, in my opinion.

This equipment looks to be in good shape—if you are looking for 8-frame equipment. If everything you have is in size 10, this equipment would drive you crazy the first season you use it. Even free, this can be a bad bargain.

At first glance, this pile of equipment appears to be mostly kindling, and without any history on disease it's a real gamble on whether it has any value at all. When weighing the asking price (if even there is one) consider the time it would take to simply evaluate all this, plus the time to paint and repair those that could be saved. If your estimates make you money, then the obvious choice is to haul, repair, and use. Too often though, it's haul, evaluate, find out it's junk, and you see money and time go up in smoke.

A long-term lease on a small piece of land that you use for storage, for a beeyard, for parking your truck, or for anything you need land for, is an uncommonly good idea. With luck you'll be able to put up a reasonable building, or just a canopy shelter. A permanent bear fence is ideal, and year-round access is important. Good forage nearby, safety from agricultural pesticides, and lots of year-round safe drinking water are also required. You have a much better chance of finding all of these things if you pay for the land, rather than hope the free space you are squatting on works.

Don't overlook portable storage containers as temporary shelter and storage on a piece of leased land. A small cement slab or just cement anchors to hold the canopy poles, a few containers, and you have a home away from home. And, perhaps, a container for fumigating equipment.

Breakthrough: Lease Land for Beeyards

Find the perfect spot—and you know what you need and want—and lease the land from the owner, or maybe from the person already leasing the land. Why not look for a long-term lease for ten good acres on which you can store equipment, which has good year-round access, available and safe nearby forage and water, and is out of everybody's way? Wouldn't that be worth the money?

Why make your life more difficult by putting up with free but second-rate yards. You get exactly what you pay for. And if you stop to think about it, how much are you paying for that "free" beeyard in honey every year?

Farmers rent or lease land to make a living. Why shouldn't beekeepers do the same?

Expanding Your Honey House

Producing more honey means you will need more storage, jars, pails, and bigger equipment to handle it all, which leads to more outlets to sell it in and more time to do it all. And that's just for the honey.

Your expansion plan must handle all these things in unison. You can't buy 1,000 colonies and handle them with your 6-frame extractor. Everything must more or less grow together, so your plan has to accommodate the rate in which you expand, and you'll expand no faster than the slowest part of your plan.

A typical situation is that more honey is the goal, or, as sometimes happens, more honey is the result of whatever you changed. More honey leads to one of two things: lots more time spent in a too-small processing facility, or upgrading the processing facility to be able to handle all that honey. (Another example of time versus money.)

- Before you begin the process of upgrading, considering equipment that works together as a team is critical.
- Figure out your handling and storage capacity before you make your first buy.
- Build that processing plant in your mind first (like our uncapping friend from earlier). This way, the last thing you buy will be sure to fit with the first thing you bought.
- Do your homework. This is a good use of time that will save you money.
- Establish a good business plan.
- Find good references on building honey houses and the right kinds of equipment to put in them.
- Talk to as many beekeepers as possible, find out what they did wrong, what they did right, what they'd do different, and what is just right. Get the answers to those questions from a dozen beekeepers, and you'll be able to determine the perfect plan for you.

Making sure all your equipment talks to each other is the goal for an efficient honey house, especially if you spend most of the harvest season doing the work yourself. If you routinely hire part-time help during harvest, the layout and patterns of movement will be different. No one should have to wait for someone else to finish her job before he can get started. There should always be more to do, rather than time to waste. The operation shown here illustrates that: One person moves full boxes to the uncapper and removes boxes with empty frames away; another removes frames from the boxes and loads them into the uncapper; a third person is cleaning frames while still in the uncapping trough and then loads them into the idle extractor; a fourth unloads the other extractor and replaces the frames into the empty supers.

When considering a new facility to accommodate your expansion, make enough room for supers; drums (both empty and full); wax processing; painting and preserving equipment; and an office to work in, perform testing, and hold meetings.

Painting boxes

Stacks of supers

Barrels

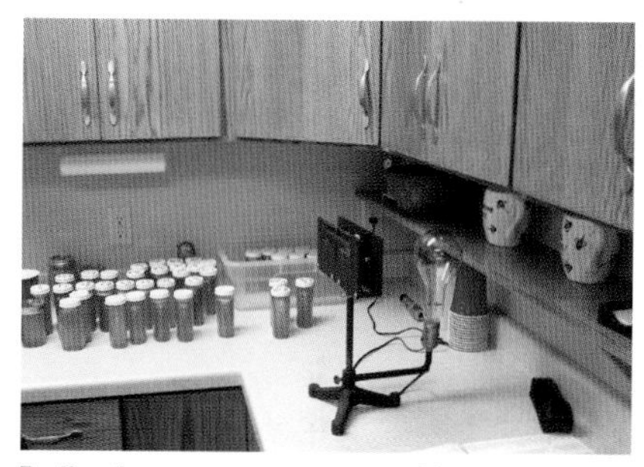

Testing honey

Pouring wax

Knowing how to hook up all the electricity, water, and heat you need can be a challenge. If you aren't sure how to do it yourself, hire an expert.

New Ways to Expand

If more of the same way of doing things is where you want to be you can turn to the next chapter. But I don't think you do. Are you really doing things the same way now as you did five or ten years ago? No, you are not. Nor will you be doing things the same ten years from now. A good plan confronts change, smacks it in the face, and takes command. You should be the agent of change in your life and in your business rather than letting the world tell you what to do, or having to conform to the wishes, the whims, and the ways of someone else. Seize the moment; take control.

More Better Honey

But wait, simply making more honey may be your goal. More hives, bigger equipment, more honey, and more beeyards, and with them the potential for different kinds of honey, especially if you actually look for beeyards with different nectar sources. Without doing things much differently, producing a varietal honey in one or more of your new yards is possible. And varietal honey is worth more than run-of-the-mill wildflower honey any day.

Several beeyards, each with a specialty, and each leased for best location, will enhance your retail line and increase the income from your honey. It should more than pay for the lease, so you get more money, and a safe and profitable beeyard in the deal. Don't settle for the beeyard you can get; get the beeyard you want.

When your beeyards share space with poisons, your bees lose, you lose, and your business loses. No pesticides are safe for bees. Insecticides kill adult and infant bees, and fungicides kill infant bees. Do not settle for a beeyard anywhere near this certain death trap.

Where are you now, where do you want to be, and how will you get there?

Summary

We've looked at where you are now, where you want to be, and even how you will get there. It's mostly about time and money and balancing costs with income. And time is both a cost and an income. If you are wondering how all this relates to the rule of rights we discussed earlier, these are the behind-the-scenes techniques and skills needed to get the right number of bees of the right age to the right place at the right time and have them in the right condition.

A good business plan or model is the yin to the yang of what that plan will accomplish. I suspect you already know what to do and how to do it and have the skills to make it work. Usually it's the equipment, or the location, or the commitment that holds us back. You need to know all of this before you begin growing your operation. Otherwise, you are always playing catch-up: too much of this, not enough of that, none of those, the wrong kind of these, and never enough time to do it all.

So, plan it out. Where are you now, where do you want to be, and how will you get there? It's simple, really.

Chapter 2:

A New Look at Honey Production,

(or If Muhammad Won't Come to the Mountain ...)

For generations, beeyards were stationary and did not need pallets, or forklifts, or straps to hold them together. They seldom moved.

For generations, beekeepers kept bees pretty much one way. With lots of trial and error they'd search for and find beeyards that fit the criteria needed and keep their bees there for as long as they produced honey and wintered well. The best locations were good for honey production from very early spring until late autumn, had access to ample water, were accessible by vehicle anytime during the year, and were (mostly) safe from agricultural pesticides. But no matter where a good beeyard was located, the greatest variable—the weather—could not be controlled. But good yards, those with a succession of blooms and minimal bad weather problems—occasional drought, late or early freezes, damage from severe storms, or other events—were sought after and guarded because making a living was the name of the game and there was seldom a season that went bust.

Occasionally, colonies were moved for off-season protection from harsh weather or winter accessibility, then moved back to those prime active-season locations when the time was right—when it was safe to catch as much of the early-season flow as possible. With well-chosen locations, beekeepers produced average to fantastic honey crops most seasons because if one of the major nectar flows failed earlier or later, other crops made up for that failure.

Sometimes, in some regions, hives are moved to better locations to ride out winter storms.

What It Once Was and Why It Changed

The variability of the weather, then, could be accommodated, if not overcome. Stability in local agriculture and other land-use planning, however, was usually more of a problem. After a three-to-five-year evaluation, locations that consistently performed poorly or had other problems—predator pressure, changing agriculture that removed honey-producing crops or increased pesticide risk, land use that made the location unacceptable such as development, de- or reforestation, or pasture growth—were generally abandoned and more favorable beeyards sought.

Meanwhile, the pace of urban development didn't slow a bit, and as cities expanded they overtook what was always the best farmland as they grew. And agriculture in general changed, easing away from the typical small family farm with a diverse collection of crops and animals, and became larger and larger and more and more specific. Farmers could once hedge their bets against any one crop or animal failing because they could depend on the others to carry them through a bad year. Once industrial farm management became the norm, only one or two crops were produced on vast acreages—efficiently, rapidly, and in typical commodity-like fashion.

As fencerows, diverse crops, roadside weeds, meadows, pastures, and fallow land disappeared, so too did profitable honey crops from those locations that had been so predictable for so long.

Cities often form near prime agricultural areas to support the outlying farmers. As these cities expand, they move into the prime farmland that supported all those farmers.

Huge, continuous tracts of land came to be, the combination of many, many smaller parcels.

Bees are now constantly on the move; palletized, mobile, and ready to go.

A monoculture crop, such as soybeans, canola, almonds, or corn, provides nectar or pollen in a deceiving abundance. A honey bee's diet with deficiencies in a few essential nutrients means that starvation in the midst of plenty has become a common disorder. This is a serious problem in areas dominated by a single agricultural crop, and acceptable alternatives do not seem to exist.

Over time, as foreign honey pressure increased, moves became even more pollination-contract driven, with honey crops only occasionally included in the trip. Sometimes honey production was in the equation; just as often, they never stopped at all.

How Beekeeping Changed

To accommodate these changes the traditional practices of working bees in season and devoting winter months to other pursuits soon went the way of those fencerows and family farms. The need for change became even more urgent as the margins of running a beekeeping business decreased due to steadily increasing operational costs while the pressures of the global honey market kept prices low. Keeping bees for less than twelve months a year disappeared because to compensate for the additional pressure beekeepers began moving bees from crop to crop, location to location, extending the season to all year long, to compensate for the additional financial pressure.

In this migratory world, colonies spend winter months in cushy locations, get divided, make honey, and maybe have a local pollination contract in the mix. As spring approaches, colonies move with the advancing season.

Agriculture has become more specific and crops needing honey bee pollination have expanded where they grow well at the expense of less profitable, and usually more diverse, small holdings. Huge areas became devoted to a single crop, and everything in the way has been removed—fencerows, woodlots, vegetable patches, meadows, and unused land—and these areas now support fewer and fewer types of plants, which has created its own set of nutritional problems for the bees that were moved there for pollination.

Huge loads of bees move from crop to crop to crop. It is industrial beekeeping at its best.

Monoculture pollination, such as this almond orchard, is deceptive because a plentiful, but narrow and restrictive, diet is not good for bees (or for people).

The assured income from sequential pollination contracts paid the way, and the incidental honey crops became a bonus. Early on, a substantial honey crop at some time during the year was often a financial necessity, and a failed crop could be a hardship. Interestingly, just as before this all began, inclement weather doesn't change the pay plan for a pollination contract, though it sure interferes with honey production during the season. Pollination income was stable. Honey production wasn't.

More Change: Honey Money

But producing honey crops and completing pollination contracts were often in competition with each other from a management objective. Good honey-producing colonies seldom make good pollination colonies due to timing, moving, and harvesting. Pollination took the lead in a business because a future pollination contract was a promise, while a future honey crop was, at best, a gamble. Pollination income became the staple of most commercial and even sideline operations, while honey production diminished in importance.

Interestingly, on occasion there would be seasons when the global honey situation was such that a good honey crop could be as lucrative as several good pollination contracts—and choosing the right direction early enough in the season was difficult. In fact, this intermittent reinforcement of good honey money on occasion trained beekeepers not to abandon honey production, no matter the more often outcome of extra work and little money. It was nearly impossible to wean beekeepers off the honey bottle, and the management struggle for many meant running colonies for both, often excelling at neither.

In the United States, eventually, a few large crops needing pollination came to dominate the business. Almonds are king, requiring far more than half of all the colonies in the country to be in California for more than a month. Larger beekeeping operations sought contracts for cranberries, wild blueberries, apples, cucumbers in some places, cherries, plums, vegetables for seeds, and forage for seed because increasing prices were the norm. Supply and demand, no matter where in the world, for pollination was important and set higher and higher contract prices.

Some of these operations, however, did make both pollination and honey production work during the year. The goal was to invest heavily in early pollination preparation and direct most of the colonies for a few early crops. After making good on early contracts, colonies were often temporarily used for other pursuits—primarily making queens and bees to replace colonies lost during pollination, but also making packages from strong colonies and selling them to distributors. This worked for swarm control, got rid of a lot of old bees, and made good money. Remaining colonies, including new splits, would go to mid- to late-season honey-producing areas for what was still a profitable crop.

Gorse is an abundant honey plant in the UK, with great expanses found. But note there is essentially nothing else blooming, and a steady diet of even good honey plants is not good for the bees because of deficiencies in some essential elements. Variety is the spice of life, and the key to a healthy diet.

Sometimes colonies were moved to honey-producing areas such as the Conservation Reserve Program (CRP) land in the state of Mississippi in the United States.

Bees are moved when it is convenient for the grower, not the bees or the beekeeper, and midnight moves are common, and dangerous.

When colony collapse disorder first became an issue, both scientists and beekeepers were surprised at what they found. One major study found that during a good season more than 50 percent and sometimes as many as 100 percent of an operation's colonies expired sometime during the traveling season.

A Price to Be Paid

But there is a price for following those blooms—whether for pollination or nectar collection—that the bees and beekeepers had to pay. Moving bees is stressful for colonies due to the quality of the equipment used to move bees, the frequency and the time between moves, the nutritional quality of the crops pollinated, the weather during and between contracts, and differences in exposure to agricultural pesticides. These situations cause a multitude of problems, including frequent queen loss and disruption of colony cohesiveness. When colony collapse disorder first became an issue, both scientists and beekeepers were surprised at what they found. One major study found that during a good season more than 50 percent and sometimes as many as 100 percent of an operation's colonies expired sometime during the traveling season, and easily two-thirds of the queens in any remaining colonies needed replacement—either by the bees, which was common, or by the beekeeper if he happened to notice a queenless colony.

Meanwhile, beekeepers become nomads, traveling between posts with schedules dictated not by a calendar but by the climate, the weather, and agricultural practices. Eventually these practices so stressed colony health and morale that some beekeepers are replacing almost 100 percent of their colonies on an annual basis. Honey bee reproductive biology allowed this, but the financial burden was increasingly problematic.

Imbalanced Nutrition

To address the needs of the growers, bees were force-fed to meet early-season strength requirements. This meant timely and expensive feeding—expensive in terms of labor and product—and pushing bees beyond their normal growth curve. And even the best substitute honey bee food was not very good. Beekeepers and bee supply companies suddenly became aware of the significant lack of good information regarding honey bee nutrition and had to scramble to make their own diets for their bees. The basics of the honey bee diet had been woefully neglected by the scientific community. And though adequate, and certainly ahead of other discoveries, supplier-produced honey bee food wasn't as good as the real thing, and improvements were slow in coming.

When combined, the additional costs to replace lost queens and whole colonies, the extra feeding, the stress of constant migration, and the never-ending, constant exposure to agricultural pesticides, made some beekeepers evaluate this lifestyle and consider alternatives. But safe and sane alternatives are few and far between. There are old-style locations that still offer an abundance of safe forage, good wintering attributes, and accessibility, and are within a reasonable traveling distance from the beekeeper's headquarters. But these locations are disappearing. Land-use practices are evolving to accommodate pavement, people, and the plow to make a modern-day monocultural desert. Some locations are not routinely exposed to agriculture pesticides for the simple reason that the area is not a good agriculture area. Typically, areas shunned by farmers are shunned deliberately.

And since the only way to be successful in industrial farming is to get as close to 100 percent efficiency as possible, crop choices are limited, which means that diversity continues to disappear in the fields and meadows of the Earth.

High-fructose corn syrup (HFCS) took the place of natural nectar and regular sugar syrup, and feeding bees the cheapest possible diet became a way of life to meet migratory demands.

When bees are placed next to the crop to be pollinated, there is seldom anything else for the bees to forage on. For many crops, growers require so many bees to do the pollination work that there may not even be enough bad food to go around to adequately feed the bees. After pollination, the colonies are in terrible shape and need additional, and artificial, food just to stay alive.

Non-Agricultural Areas

Monoculture prisons and destructive developments are not the only choices. For instance, some tropical and near-tropical locations support nearly continuous plant growth with a succession of naturally occurring bee forage. Honey bee colonies have the opportunity, then, to produce honey nearly continuously. But even here seasonal changes interfere with continuous bloom. Plus, these tropical locations usually have a downside: continuous predation by pests and predators.

If you want to survive in the tropics you have to be tough. Witness the success of African honey bees, and the failure of (traditionally called) European honey bees in the tropical areas of South and Central America and southern North America. Though fantastically successful as an unmanaged but exploited tropical inhabitant, the bees from Africa have proved largely unworkable from a commercial sense in the Americas in any of these areas, while European bees could not successfully compete in the tropics. New-world tropical and semitropical areas have become largely second-class production areas from an industrial sense and with few exceptions have little effect on global honey production.

Honey bees living in more moderate (nontropical) locales experience more clearly defined seasonal activity and little or no interference from the African honey bee. In more moderate regions there is a time to blossom, a time to ripen, and a time to rest.

The ripening and resting times are unproductive when making honey or requiring pollination services. If other varieties of local flora are not naturally providing bloom on a nearly continuous basis, beekeepers have to find other, probably agricultural, bloom or move.

Areas in the UK where the heathers occur naturally have been expanded so there are more plants available for the bees to visit, and more crop to harvest at the end of bloom. But the bloom time itself is not lengthened.

Citrus groves are used by beekeepers to extend their season in the fall and to make honey and expand their colony numbers early in the season, before other nectar crops begin producing. This practice is changing in the southeastern United States because of citrus disease and constant pesticide applications. Orange blossom honey is slowly becoming extinct.

Expanding Natural Bloom

Growers can increase the amount of bloom of naturally occurring forage by allowing what exists to expand. (Wild blueberries in Maine in the United States or heather in the UK are examples of natural expansion by removing competing vegetation.) This increases the potential honey crop gathered from those plants during the time of bloom, but doesn't increase the length of time these plants are blooming. Therefore, the volume of that particular honey crop may increase, but the bloom time remains the same. A beekeeper with many colonies can capitalize on this situation because he can accommodate the larger scale of the operation.

But why not find ways to lengthen the time of bloom?

Early Nectar Crops, Then Farm Crops

Bees and beekeepers are subject to the seasons: In the northern hemisphere the closer you are to the equator, the earlier in the calendar year the blooming season begins, and the earlier the blooming season ends (in most places). For example, in the southern United States trees and shrubs in the maple family (*Acer* spp.) and willow family (*Salix* spp.) may bloom in mid to late January, but these and the naturally occurring uncultivated nectar and pollen sources that follow are finished by the end of June or early July. From then until the next flush of bloom, which may be several months away, the bees are on hold, either eating their stored food or being fed. These are typically called welfare bees.

To modify this scenario, beekeepers capitalize on locations where agriculture partially fills that bloom gap. When the earlier, naturally occurring nectar crops are close to finishing, agricultural crop nectar flows take over, thus extending the season. Agricultural crops that do not require pollination such as citrus, cotton, and legume forages can provide a nectar crop for some time afterward. But when those crops are done the dearth begins again in earnest.

Migratory Beekeeping for Nectar Crops

After the early crops are finished, bees and beekeepers can migrate to find plants that are blooming, most likely in areas far from the original location. This is customary, and migratory beekeeping needs a book's worth of management instructions. Migratory beekeeping reduces or eliminates the downtime and welfare bees that a colony experiences in a dearth in one location, but invites a whole set of situations that, frankly, many beekeepers want to avoid.

Are the days of plenty in the almond orchards coming to an end? Renting bees is expensive and almond growers are diligently searching for self-pollinating varieties to take away the sting of pollination rental fees.

Many beekeepers want to be done with the headaches of migratory beekeeping because the more places bees go, the greater their chances of colliding with trouble. When the move is to agricultural crops the greatest threat is their exposure to pesticides at levels that can be both immediately life-threatening and long-term chronic or sublethal.

Any time bees travel, their chances of mingling with other beekeepers' bees increases. And how do you know if these bees are safe to associate with?

Studies of colony decline and collapse demonstrated that viruses are transferred from bee-to-bee in the hive and from bee-to-flower-to-bee outside the hive. This should give caution to anyone moving next to unknown bees. And moving from the original nonproductive area to a monocultural crop area won't solve a food shortage because the new food may be even worse for the bees than no food at all. And even if it's not, it's only a short-term solution.

In a beekeeping world dominated by global markets and slim margins, seldom does honey production on a short-season crop generate enough of an income to sustain a beekeeping business. The added financial benefits of several pollination contracts during the year have devolved from being benefits to being business-sustaining necessities. And, right at the peak of this frenzy of how to run a beekeeping business, the almond industry discovered the genetics of self-pollinating blossoms, and quite suddenly this becomes a very different story.

Over the course of the next decade, it is imagined, the almond industry will slowly adopt this cost-saving benefit, reducing (but, as with seedless watermelon, not eliminating) the need for pollinating bees. The change won't be instant, but it will change, and honey bees and almond blossoms will no longer be the marriage made in heaven. This is a prime example of how, without the additional income from pollination, a business could not long sustain itself on just a single honey crop.

Summarizing the Dangers of Moving

Short-season honey crops generally produce too little honey to adequately support a business. You can supplement early crops and extend the season if there are nearby row-crops, but you add in the dangers of pesticide exposure and a significant lack of dietary diversity. If you move your bees to new crops to extend the season, you have moving stress and expense, even if the crop is productive. If you are moving to anything agricultural for pollination income, you face exposure to unknown bees with problems you don't want, additional exposure to agricultural pesticides you don't want, and reduced dietary diversity that you don't want either.

Planning a Honey Crop

So, What's a Beekeeper To Do?
Honey crops can be planned, planted, and harvested on a schedule and in locations that work for both bees and beekeepers. A safe but planned nectar crop that blooms after local natural flora bloom is finished is a benefit. For instance, think of a 100-acre (40 ha) fallow field with a crop in flower, but at a density that's almost nonexistent. Then concentrate and expand that planting so it contains more than enough flowers to be valuable as a honey bee forage location.

These wildflowers could have additional benefits. For instance, they could be a harvestable forage crop to feed cattle, produce wildflower seeds for sale, produce herbs for oil or processing, provide green manure for the soil, or be used to remove excess nutrients or moisture from the location. Expanding on this concept, additional plantings could be more perennial in nature—trees or shrubs that bloom before or after other intentionally produced crops. These, too, could have additional value, such as timber or other income-producing material.

To produce these kinds of honey crops requires a whole set of management skills and equipment most beekeepers don't routinely have at hand. Even if they did, they seldom have time to devote to farming, given their commitment to their beekeeping business. Essentially, to produce these secondary crops, a beekeeper needs to also be a farmer with the machinery required for preparing soil; applying fertilizer or manure; planting; controlling pests; disease, and weeds; harvesting; storage; and marketing.

That's a lot to ask for. Maybe we don't have to.

Planting a wildflower mix in an empty space between permanent crops—such as these grapevines (left) or on otherwise unused land between structures such as these greenhouses (above)—offers pollinator forage in an otherwise desert of abundance.

Before the turn of the last century A. I. Root was searching for answers to more and better honey plants and inexpensive production techniques. But still today, owning all the needed farm equipment to raise honey crops would be an expensive option.

White clover (top) is ideal as honey bee forage and soil builder, and makes an excellent hay crop. Vetches, too, are excellent for both pollinator forage and hay crops (above).

Forgotten or neglected spots along driveways and field roads where machinery typically can't reach are perfect for plantings.

Types of Honey Crops

1. Annuals that require minimal inputs in terms of soil preparation, planting, seed and fertilizer costs, and crop protection.
These types of plantings cost little to get going and can produce a bountiful honey crop in a short time, but usually produce little or no value-added crop. This may be by choice, if there is no way to process the crop in the immediate area, for instance. Buckwheat for green manure production or to dry out wet soils rather than as a seed crop may fit this model. These types of plantings can be beneficial to a grower, the land, and the beekeeper, even when time and equipment are limited. Some, however, may be abundant self-seeders, so future crops may be produced at no additional cost. These often slowly decline in vigor and abundance, but a minimal investment can produce an exceptional return.

2. Ground cover cash crops.
Forage, for instance. Any of the many, many legumes used for soil improvement or animal forage can be left to blossom before being cut and baled or chopped. Many will bloom two, three, sometimes more times in a season. These cover crops have great value for bees, beekeepers, and farmers and should not be overlooked. They can be bread-and-butter crops for this sort of honey-producing enterprise, and they have additional marketable value after harvesting. Plus, they tend to be short-time perennials, so replanting is minimized to every four or five years. Moreover, plantings can be scheduled so there are blooming crops from very early to very late in the season.

3. Establish a crop of perennial, native wildflowers, shrubs, and even trees.
This level of planting is possible, if land, time, and equipment are available. This can take time, specialized equipment, and some basic skills, but the end result is a continuously blooming assortment of flowers contributing to the success of nearby colonies. There are already government-supported programs in the United States that support land preparation, seed, and plant purchases and the labor to get them installed. This type of planting is usually the fallback for colonies when agricultural crops are not blooming. They fill the gaps

and keep the bees going between major flows, whether planned or naturally occurring.

Good examples of these plantings include roadsides, stream banks, fence and hedgerows, and the like. There are, of course, other agriculture crops that can do the same thing—some that produce a value-added crop, others that are provided for blooms only. A profitable attribute of this mixed planting scheme is the diversity of diet it provides its pollinators (honey bees and others, of course), and the reduced chance that a pest will become very destructive because there are many varieties of plants, only some of which are targets for that pest.

Several groups concerned with just this sort of problem are actively promoting these types of plantings because they have mixed in-bloom dates, pest pressure, soil and water use, and attractiveness to a wide range of pollinators for food, and wildlife for habitat. And, especially, if government supported, they are all well adapted native plants.

Larger areas, normally left alone because of terrain or accessibility, occasional flooding, or other reasons can be used for several varieties of bee- and wildlife-friendly plantings.

Plantings Help More Than Just Bees

Bee forage areas also provide habitat and refuge for migratory insects and pollinators such as butterflies.

These habitats also provide protection, nesting areas, and forage for wildlife, which help with natural pest population control.

Pollinator protection is definitely a plus when establishing these plantings, providing pesticide-free areas that offer good nutrition and diversity of diet and habitat.

Maintaining a balance of pests and predators for the plantings is important.

Planting Perennial Shrubs and Bushes

There are a host of perennial shrubs that can be planted for a honey crop that should be considered in this scheme because they establish bloom faster than permanent trees do, while providing abundant bloom density in a compact space. Most of these are long lasting and produce both nectar and pollen. Some conservation organizations provide inexpensive seedlings for plantings. These can be planted in great numbers, and while expecting some percent loss over time, an acceptable number of plants reach blooming age.

A good example of this is planting honeysuckle (*Lonicera* spp.) shrubs. A good estimate of planting requirements is to provide 300 one-year-old seedlings to cover an acre (0.4 ha) of ground, whether a solid acre, or spread out along fencerows, stream banks, or in hedgerows. Generally, 200 will make it to maturity in three to five years, providing an abundant early-season honey crop for a reasonable cost, plus a couple days' labor.

Honeysuckles make excellent forage, are long lasting, spread magnificently, require essentially no care, and are simply loved by honey bees.

Willows bloom very early in the spring, providing excellent "real" food to propel colony growth when it is needed most.

Planting 300 honeysuckle seedlings anywhere equals an acre's (0.4 ha) worth of planting. You don't need a square acre of land to produce a square acre's (4,000 square m) worth of honey (for the two weeks' worth of bloom you'll get). Remember that.

A planting of crab apples offers both landscaping beauty and an ample amount of food in mid-spring.

Long-Term, Permanent Plantings

And finally, consider long-term permanent crops—primarily trees. Careful selection can provide spectacular bloom density from late winter to late summer in abundance, with minimal land use. Of course, there is the time lag between planting and nectar and pollen harvest, but with careful planning some trees bloom even the first year, and are in full bloom after two or three years, filling in for later maturing trees that may require four, five, or more years before producing ample bloom. It takes a time commitment to invest in this crop, but the financial burden is minimal and the payoff is generally profitable and generous.

Planning for these plantings should be similar to those of perennial shrubs—overplant inexpensive seedlings with the expectation of losing a third or more before reaching maturity. So for most full-size trees (size here is relative, since many arid land trees generally don't attain the size and stature of most temperate woodland trees) plan on planting at least twice as many as you'll finally need—maybe as many as three times. For a stand of tulip poplars, for instance, plan on planting 100 for an acre's (0.4 ha) worth of bloom, expecting maybe 40 or 50 to make it to maturity.

Growing Your Own, with Help

This concept has several names—I've always referred to it as land-based honey production, but the name isn't important. It can take much of the gamble out of making a honey crop. Planned and planted correctly, colonies managed the right way for this scheme will produce honey from early, early spring until late, late autumn without having to move those colonies a single inch. The benefits of such a plan, I trust you can imagine, are numerous.

There are thousands of details that must be addressed to make this profitable for bees, beekeepers, and farmers. But there are some insurance factors that can be installed that will spread out the risk of even a partial crop failure. That's because there are many overlapping crops during the season. One plant's failure isn't the end of the season and it won't place your bees in a starvation situation because there are redundancies built into a well-planned scheme. This type of operation brings us back to the days of many small, diverse farms spread out over a large area, rather than one huge corn, soybean, or wheat field.

Of course, the reasons beekeepers moved away from stable and permanent locations haven't gone away. Honey as a crop is often still a commodity. And if it is produced like a commodity it will command a commodity price even if the plantings produce well. However, if the varieties of plants have relatively well-timed, sequential, and overlapping blooming periods, and there are enough of them to produce an average—or better, an abundant—crop, then producing varietal, or at least well-timed artisan, honeys is possible. With all this well-timed artisan honey comes added value and more income. And still no moving, no pesticides, and only good food to eat.

Moving bees takes time and money and extra equipment, and getting your truck or trailer stuck in the middle of the night in the middle of an orchard is no fun.

A truck accident with bees always makes the national news, and not in a good way. Though insurance will cover the repair costs of the truck and the hives, it doesn't cover the bees that are lost.

Other Cost Considerations

And while we are considering income, don't forget to deduct the previous costs of simply moving bees:

- Loading and unloading and moving and setting colonies.
- Queen loss.
- Beekeeper upkeep and windshield time.
- Trucking costs (and accidents).
- Bee nutrition: getting bees to thrive on the nutritionally challenged monoculture crop they would otherwise be subjected to.
- Exposure to bees you might not otherwise want to know.
- Exposure to the pervasive pesticides bees encounter in the real world of modern agriculture.

Today's hard-core commercial beekeepers will, during an average year, replace as many as 80 to 100 percent of their colonies each season. This may be an acceptable business plan—acceptable, but certainly not sustainable. Remove these costs and suddenly the income from producing stay-at-home varietal and artisan honeys begins to look promising.

This isn't an agronomy or a horticulture text, nor will it supplement for one. If you are to be the producer of these crops you need to consult a higher authority and more informed resource than you will find here. The intent is to provide enough of the information you need to ask the right questions. However, the increasing threat of losing pollinator habitat, including honey bees, has produced a variety of these very resources available from local, state, and federal government agencies in the United States and comparable institutions in other countries. For the most part they are free, and easily available on the Internet.

How Much Land is Required?

The least understood link in this program is the amount of land required. It takes a sizable chunk of space to support a large number of colonies. A backyard garden isn't going to do it. A current, and probably pretty good, estimate is that it takes a continuously blooming acre (0.4 ha) of blossoms to support a colony for a season, but this needs further examination.

For instance, early in the growing season, even though you are urging your colonies to grow fast to accommodate the major nectar flows later in the season, the population of the colony is smaller than it will be later, so the estimate of an acre (0.4 ha) is a bit overstated. However, I suspect that at the height of the season when colony population is peaking, that an acre (0.4 ha) may offer only just enough for a thriving colony, so that acre requirement may be just a bit understated.

A good rule of thumb I've heard is that you'll need 1.25 blooming acres (0.5 ha) all the time—from a month before last frost to a month after first frost—to support a single colony. That allows for some downtime between blooms due to erratic weather or other climate-related issues.

Considering this scenario, and knowing that continuous bloom is required to keep growing colonies growing, the blooms doing the feeding can be spread out over acres, or clustered together. A good example would be a tree stand of, say, an acre (0.4 ha).

Simultaneous bloom: Bees from some colonies will be working the dandelions on this orchard floor, while bees from other colonies will be working the apple blossoms above. Some will be working both.

Calculating Foliage Density

First, look at the area that is covered by the foliage of a tree—here, say a circle with a diameter of 40 feet (12 m). The area covered by that tree then is 1,250 square feet (116 square m). About thirty-five trees would comfortably cover an acre (0.4 ha). But consider the density of bloom on a linden (*Tilia* spp.) tree, for example, compared to the density of bloom of an acre (0.4 ha) of almost any ground cover. When a tree blooms, observe that the blossoms are essentially stacked one upon another. The same goes for almost any tree, and, to a lesser degree, any of the perennial shrubs you might consider planting.

This type of density calculation even applies somewhat to some ground covers. Dutch clover, for instance, produces several blossoms per plant, and if planted at recommended rates will produce in the same land area as that tree more than a 1,000 plants with more than 10,000 blossoms. An anecdotal figure handed down to me by Roger Morse, the apiculturist at Cornell University for years, was that an acre (0.4 ha) of trees produces a million blossoms. With newer varieties of trees, I suspect this is now a very conservative estimate, but it's the only one I have, so we'll use it for now. This calculates to nearly 29,000 blossoms per bloom cycle for a linden tree, compared to the optimistic 10,000 blossoms of the clovers planted in the same area, right beneath that tree. Clearly, floral density during bloom on that acre has to be considered when measuring the production of any potential planting.

Another consideration is simultaneous bloom. For instance, under that canopy of linden bloom there could be a ground cover blooming at precisely the same time, which lends itself to further examination in total production area. If a colony requires about an acre (0.4 ha) of bloom to sustain itself, then if on any given acre there are two, perhaps even more, blooming events ongoing—at ground level, a couple of feet above ground level, and 30 feet (9.1 m) above ground level—does this essentially double or triple the production of that space? Absolutely it does if the density of bees is such that all blooming events can be taken advantage of. Therefore, your acre of thirty-five linden trees, which by itself, because of the density of bloom was producing at more than an average acre (0.4 ha) anyway, now, with this additional bloom, is producing as much as 3 acres' (1.2 ha) worth of food concurrently.

These considerations need to be both planned for by necessity and taken advantage of when possible. It can become a complicated, sophisticated program that constantly needs tweaking as different pieces of land come and go in the scheme of things. It is, however, less complicated than finding new beeyards every year, working out pollination contracts and making grade for every crop, or promising a boatload of queens or a truckload of packages by a certain date regardless of the weather. And it's certainly safer than letting your bees wander about.

Foraging Behavior

You'll recall that a colony rarely invests its entire foraging force on a single nectar or pollen source at any one time. There may be a 50-acre (20 ha) patch of something flowering right next door, but some percentage of the colony population will be heading off in the opposite direction to gather an altogether different food for quality, availability, or quantity.

Nevertheless, when combined, the area covered by all the foragers at a time will approach an acre (0.4 ha), or some portion of an acre based on the colony's population.

It's hard to imagine where available land is until you see it from 30,000 feet (9.1 km), or use a computer map program to show where land is or where you want it to be. In fact, these programs can make finding available land much easier, so don't neglect to use them..

Working with Landowners with Lots of Land

Landowners too often overlook the income opportunities that providing food and habitat for pollinators and wildlife offer. Your job, should you choose to accept it, is to find those landowners and convince them of the benefits of supplying land for you, your bees, and a host of other pollinators in the process. The upfront costs of turning even a small plot of unused land into a bed and breakfast for your honey bees cannot be overlooked. Seed costs vary, and finding a supplier that has competitively priced seeds, and is close enough that the freight costs don't eat up the low pricing, should be straightforward. You will need to strike a bargain with the landowner, and it may involve you buying the seed, the gas for transportation, and his time, and he'll get the seed in the ground—maybe 10 acres (4 ha) of white Dutch clover, for example.

Costs per Acre

Costs and income for a crop such as white Dutch clover are easy to calculate because seed costs are relatively stable. Also, because it is a common honey plant, the honey income per acre is also well known. In an average year, with adequate moisture, planted in a location with the right soil, with mowing two or three times, ten colonies can make 60 to 200 pounds (27 to 91 kg) of honey each in about three months, just from access to that 10-acre (4 ha) plot. So a poor year will bring in about 600 pounds (272 kg) and a very good year 2,000 pounds (907 kg). And that's just from the clover crop. The price of honey may vary, but even a 600-pound (272 kg) year is going to cover expenses, and an average year will double your investment.

If that 10-acre (4 ha) plot only lasts for the period of bloom, it's turned a profit. If it lasts longer (two or three years) before a replanting is required, then the profit generated becomes substantial. The land is still profitable even when you pay the farmer for the privilege of using his land for the full duration.

This sort of informal arrangement can become more formal if the landowner is participating in any of several government-sponsored conservation or pollinator protection programs. To qualify, the landowner must comply with regulations and produce approved, pollinator-friendly plants. Some of these programs are becoming very beneficial for both landowners and beekeepers due to the expansion of payments and the increase in planting diversity allowed.

If you can keep several of these little pockets of subcontracted crops running every year, your future as a beekeeper or honey producer is nearly guaranteed. And instead of looking for better beeyards, you are spending time looking for land that allows you to plant forage that your bees can maximize. You'll need several of these plots every season within flight range of your bees (who are safe at home) to ensure against crop failure in any one location, and to offer your bees several choices.

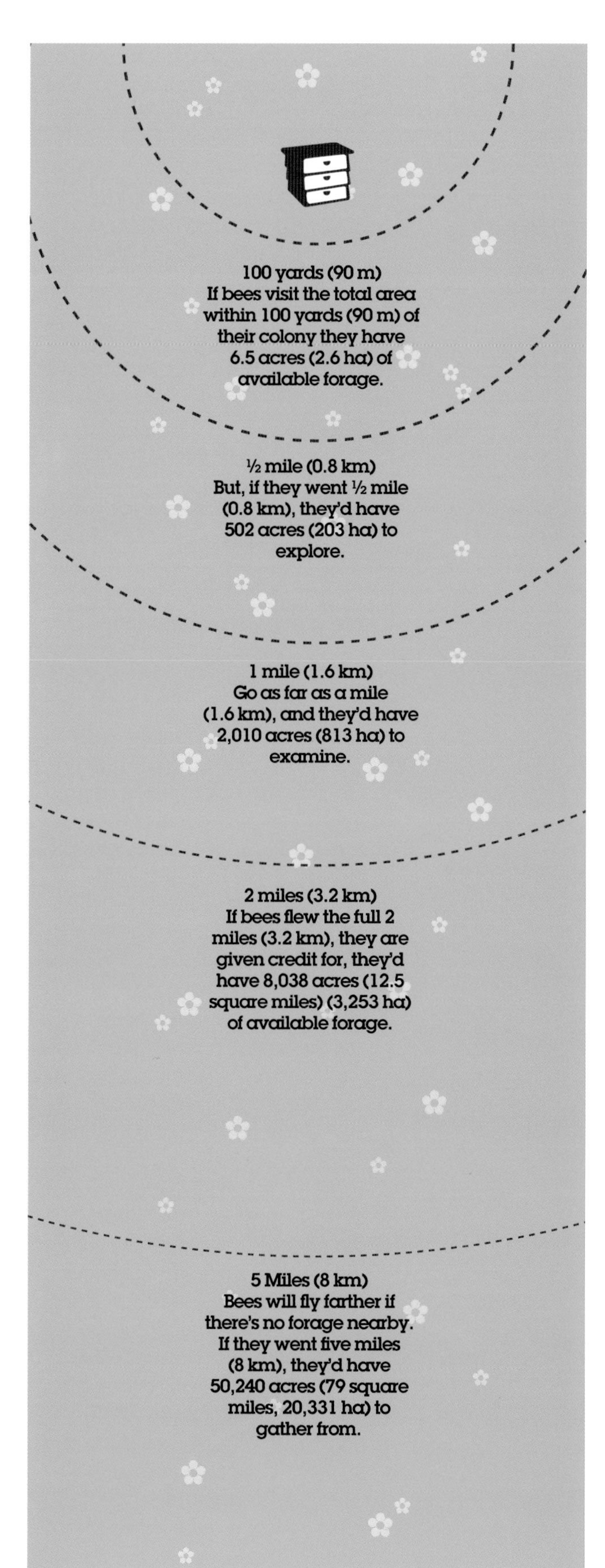

But, what about the organic crowd? Producing organic honey is nearly impossible in almost every location.

More Land: Bigger, Better, More Complicated

But even larger swaths of land otherwise not used are available: land in crop rotation programs, government set-aside programs, and cropland actually being used for crops. And there's gold in those large pieces of land, even if you don't get to use them directly. With good planning, even heavily used cropland can be partially converted to honey cropland by the owner by plantings on fencerows, field borders, stream bank areas not good for other crops, erosion-control swales, edges of livestock pastures, road edges, and wetlands. Plus, some of that land might be available for lease to people just like you.

Remember that these crops and the land they are on do not need to be contiguous. They can actually be several miles apart: a stream bed a mile west of the bees, a legume forage field next door, hundreds of trees on a string of fencerows, an agricultural crop on the next farm, and a continuously blooming meadow just over the horizon. It's the distance from your bees that's important, not the distance between blooming patches.

GMOs, Organics, Pesticides, and Other Flies in the Ointment

How do you feel about your bees foraging on crops that have been genetically modified? There are all sorts of those crops out there now, and more on the way. As of this writing, genetically modified plants seem to be a nonissue relative to honey bee safety and honey quality. I am not supporting the use of these plants, but I'm not condemning their use either. It matters less what I recommend and more what your customers demand. If you live in an area where your bees will visit the multitude of crops that take advantage of this technology, what's a beekeeper to do?

Consider that, as this is written, corn, tomato, potato, cotton, squash, papaya, rice, canola, sugar beet, flax, radish, soybean, cantaloupe, alfalfa, and poplars are major crops grown that have some level of genetic modification. They are either herbicide-resistant, insect-resistant, virus-resistant, or some combination of these attributes.

Of these, corn, cotton, squash, canola, flax, radish (for seed), soybean, cantaloupe, and alfalfa are commonly visited by honey bees. Further, honey bees are often the insect of choice for those crops needing pollination. This subject will continue to become more complicated. My advice is to ignore it as a concern, but stay tuned. In the meantime, do not let your market strategy be blindsided by this new world of agriculture; however, understand that often there are few alternatives.

There may be small organic farms that offer a safe haven for your bees, but unless the organic farm is miles wide your bees will most certainly visit other locations that are not organic: roadsides, traditional agricultural farms, golf courses, cities, suburban developments—the list goes on. Finding an organic location is certainly safer than many locations, so the extra expense of purchasing sensitive seed for this operation may be worth the price. But unless you are so isolated that your bees absolutely cannot visit anything but organic land, consider the cost to supply those seeds.

What about pesticides or an IPM program that affects your honey crop? If there's a weevil eating your farmer's alfalfa crop, and there's only one way to stop the weevil—spraying—who has the problem: the farmer, or you? It is obvious it pays to ask all these questions before you buy the seed and the farmer harrows it in. All sorts of dead bees in front of your hives because of a simple misunderstanding, and no contract, is a very expensive mistake.

Planting early-blooming redbud trees along steams and roadsides provides wildlife habitat, shade, and excellent spring forage. Plus, it's a source of excellent honey untainted with pesticides, herbicides, or genetic modification.

Recommended Plantings for Honey Bees

There are occasions, when only a handful of choices exactly meet the needs of the honey bees, beekeepers, and landowner. When a landowner has a temporary vacant spot, such as between crops, for overwinter soil erosion control, or a new bed that is being prepared but not yet ready to be planted, there's always a way to plant and use ground such as this. A wildflower mix may be a good choice for this, or a legume cover crop works—one of the many annual clovers affords excellent soil improvement, provides erosion control, is attractive, and provides both food and habitat for your bees and wildlife. Other types of crops should be considered, too: green manure crops, those that remove excess water from the soil, nitrogen enhancers—the list goes on. There are a multitude of lists available for your location that itemize quick-grow crops, ground covers, perennials, and permanent crops. These lists, which are compiled by people with a devoted interest in the subject, do not list invasive species.

These lists are far from complete. I include them here because they are familiar and easily sourced for the most part, and there is information on honey production available. Consult local Extension resources, and absolutely the Web pages of the North American Pollinator Protection Campaign for a more comprehensive guide to plants. (Also, there is an extensive discussion of honey plants and varietal honeys in my book, *The Backyard Beekeeper's Honey Handbook*, Quarry Books, 2009.)

My lists cover mostly moderate climate areas. Information for tropical and desert plantings is extremely specialized, and requires more detail than is suited for this book. (Plus, African honey bees are present in many of these areas, which precludes extensive commercial-scale beekeeping.) One organization that has collected and studied both pollination of tropical plants and planting for honey production in the tropics, plus information on agroforestry is Bees for Development, which is based in the UK. Contact information for this group is available on the Web. If you are involved with tropical agriculture on any level I encourage you to become familiar with its services.

Lists of plants to consider for honey bees, and other pollinators, are available from many nonprofit and government agencies concerned with providing both habitat and natural food for these insects.

Shrubs and Bushes

Russian and autumn olives, sumac, vitex, *barberries* spp., buckthorn, cotoneaster, snowberry, privet, blackberry, raspberry, *willows* spp., pepperbush, palmetto, gallberry, honeysuckles, sages, manzanita, titi, clethra, catclaw, golden rain tree, mesquite

Palmettos produce a fine honey, and grow in the subtropics where heat, humidity, and no freezing weather dominate.

Opuntias can form livestock-defying hedges and become somewhat dangerous when they get large, but when they bloom, honey bees are all over them.

Any of the cane berry plants—raspberries, blackberries, black raspberries—produce a delicately flavored honey, but can become a nuisance if left unchecked.

Cover Crops

Dutch clover field

Bird's-foot trefoil

White sweet clover

Red clover (shown here) Ball clover, Persian clover, alsike clover, white Dutch clover, white and yellow sweet clovers, alfalfa, lespedeza, trefoil, hairy vetch, crimson clover, sanfoin

Wildflowers

Knotweeds and smartweeds (*polygonum* spp.), yucca, borage (this is also a crop), buckwheat, wild sunflowers, asters, goldenrods, mustards (including the canola crop), heathers, milkweeds, monardas, poppies, wild buckwheat, Spanish needle, knapweed, golden honey plant, hyssops, lamium, lavender, lupines, mints, sages, fireweed.

A field of asters such as this is a gold mine late in the season.

White-flowered varieties of September aster are most common, but purple-flowered varieties are easily grown from seed.

Goldenrods are excellent fall blooming plants, are easily grown from seed, and provide excellent habitat for many kinds of wildlife.

All of the monardas are productive but can be difficult to establish.

Members of the polygonum family of weeds produce excellent honeys, some as red as beet juice, others as clear as water; all of them are exceptional producers.

There are many varieties of mustards, some blooming very early in the spring, others waiting until mid to late summer. All are favored by honey bees, and all produce nectar that granulates rapidly. Some mustards, such as canola and camelina, are produced in millions of acres and provide substantial honey crops for nearby beekeepers.

Fireweed is an opportunist, coming on strong after a forest fire or logging event and lasting until new tree growth shades it out, but for several years it produces a wonderful honey in great quantities.

Permanent Plantings

Black locust, Siberian pea tree, basswood, catalpa, cherry, crab apples, buckeyes, hawthorn, maples, sourwood, eucalyptus, redbud, acacias, holly, mesquite, palms, paloverde, tamarisk, tulip poplar, willows (tree spp. and shrub spp.), alders, amelanchier spp., tallow tree, plums, birches and poplars, chokecherry, mangrove, bee bee tree, madrone, tupelo.

You may have different species of plants that do well where you are, but might not grow 10 miles (16 km) down the road. I encourage you to experiment with some types of plants, question other beekeepers in the area, and explore your region fully before making a commitment and spending lots of money. Know the downsides of every plant you have to spend money on, plus what you can expect when it does well.

Maples can always be counted on for an early spring crop and for a maple syrup crop if they are planted with modern sap-harvesting techniques in mind.

Wild cherry trees bloom early, when almost nothing else is producing, and their color and timing make them stand out in a forest, an easy beacon for hungry honey bees.

The basswoods, or lindens, are sought after by beekeepers because of the amount and flavor of the honey produced, plus they are hardy and can grow in almost any location that has adequate drainage. The small-leafed varieties produce more and better honey than the larger-leafed varieties.

In almost any arid climate members of the mesquite family thrive and produce a honey crop, no matter how dry or hot it gets.

This Land Is My Land, This Land Is Your Land

Generally, beekeepers don't own a lot of land for planting. Keeping bees is one of the few agricultural pursuits that you can undertake where you are livestock rich but land poor. Without land, though, your options are limited by where you can safely keep your bees.

Your choices are:

- Average to poor honey crops when you place bees where you can, rather than where you'd like, and when you keep them near safe, but mediocre locations.
- Move them to pollination crops or other honey production areas but become exposed to chemical, nutritional, pest or disease stresses that you are trying to avoid.

So what's a beekeeper to do? Tenant farming has been a way of life for as long as people have been farming. The basic definition is: someone owns the land, someone else works the land. The work usually means preparing the soil; planting, tending, and harvesting a crop; maintaining the land, woodlands, and pastures; and occasionally caring for and housing livestock of various types. At season's end the tenant (or renter, sharecropper, or sometimes in legal terms, a lessee) sells the crops that were grown on the land and then shares (a polite term here) the money from the sale of the crops with the owner of the land.

In return, the tenant may live on the land, or sublease part of the

Here, oats were planted at the same time as the sweet clover between the rows. The oats will grow and mature this year, and the sweet clover, a biennial, will come back and bloom next year.

land to others so they pay him for the use of the land; use some of the landowner's farm equipment; and share costs of seed, fertilizer, chemicals, and other routine crop and livestock production costs. All manner of variations exist, but mostly, it's about renting land, raising a crop, and paying for the opportunity to use the land with the profits from the crop. Depending on the arrangement there may be an actual profit, or the additional income above expenses may be applied to a rent-to-own arrangement or some other financial settlement.

The details of leasing or renting land are almost always arranged within the guidelines of a contract. Each general type of agreement is different enough that each should be examined in detail.

Beekeepers are often exposed to contracts when they are in the pollination business. More beekeepers do not use contracts than do. Usually, it's a handshake that settles the deal: I'll deliver this many colonies of this strength on this date, and you'll pay me this much for each of them. Shake.

The number of things that can go wrong with this sort of arrangement are frightening on so many levels.

But this is coming to an end, as fewer and fewer can afford to be taken advantage of, whether beekeeper, grower, or pollination broker. Contracts have finally come to the world of beekeeping. Would you jeopardize your future, your bees, and your ability to make a living on a handshake?

Leasing land to produce a honey crop is relatively uncommon. It is so rare that nobody I know, other than me, has heard of it being done before. But beekeepers tend to keep good ideas to themselves sometimes, so this concept may be more common than I know. And, I do know some beekeepers who own enough land to grow specialty crops for varietal honey.

I suggest you explore some of the aspects of a typical land lease contract: You can download several templates from the Internet for free, or nearly so. With legal information you usually get what you pay for. An uncommonly good idea is to hire an attorney experienced in leasing agricultural land to examine any contract offered, or to draw up a new one because your situation is significantly different than most. And keep that attorney in the loop, no matter what.

Types of Contracts

Cash Lease: The lessee pays for leasing the land at a set price per acre. The tenant makes all the decisions, keeps all the crops, and owns the crop from the time it is planted until sold. The advantages for the owner are clear: There's little risk, especially if payment provisions are secure. For the tenant, the disadvantages are the payment rate is renegotiated annually, and depending on the crop, production cost may be high and the income low due to market fluctuations. The risk tends to be one-sided.

Crop Share Lease: Some percent of the crop yield goes to each—land owner and tenant. What percent depends on the contributions each makes. Both share in management decisions and responsibilities. The rent share is usually half to a third of the crop. The tenant may own the crop, or only own it after it is planted, or never own it. The owner or the tenant's interest is attached after the crop is planted so either may sell his share. The advantages are that risks are shared and the owner is involved, but how costs are shared can be troublesome.

Would you jeopardize your future, your bees, and your ability to make a living on a handshake?

This leaves the unanswered question: Which crop?

If, for instance, a forage crop is planted—say alfalfa—who pays for the seed, planting, and the rest, and who benefits from the baled hay later? Add to that, what about the honey crop? Is this, too, shared?

And what if there's a crop failure—say there's a forty-day biblical rain and no honey is harvested—but there's still a hay crop?

Or who gets the wood if you buy the trees? Or the honey if the landowner buys the trees? Do landowners reap the benefits of government payment if you split the costs of producing a pollinator-friendly hedgerow?

Dividing costs and benefits can be tricky. Do landowners have some rights to a portion of the honey, wax, and any queens produced, whether or not the bees actually sit on the land or on other land?

When putting a contract together the details can be as complex, or as simple, as you want. You just need to know what you want and what you can afford at the end of the day and the end of the season.

Other things to consider are the quality of the land, the value of contributions made by each party, the bargaining position of each party, and the overall program of the owner relative to pollinator support.

Evaluating Land Quality

Often a major consideration in leasing costs is the quality of the land. This includes natural fertility—organic matter, soil structure, drainage, flooding potential, and the like—plus the results of actual soil tests. It also includes crop history—what's been grown previously. Some crops can't follow some crops because of disease residue or herbicide history.

But basic beeyard provisions are important if your bees will be on the land. Year-round access for all manner of vehicles, water, safety from vandals, gates—everything you look for in a beeyard—should be reviewed.

Rental rates may be inexpensive because the fields are small and irregular. These are less efficient for large equipment but may fit what you have in mind.

Flooding, even the 100-year kind, may be an issue—not for the land or crop so much as for the bees floating away. Get a report on flooding history if you can.

One aspect of land quality to be aware of is the presence of, or introduction of, obnoxious, plant species, such as kudzu vines.

Erosion control is necessary, whether it's your land or not.

What happens to your crop and that easy access when there's that 100-year storm? Before the flood, there were beehives in this field.

Who is responsible for maintaining irrigation equipment and paying for the fuel or power?

It should be obvious that a leasing arrangement for growing honey plants, and perhaps providing a beeyard location for hundreds of colonies, does not normally fall within the borders of a generic land lease contract.

This needs a thorough review before planting is undertaken. Here are some general questions to include, in no particular order:

- What are provisions for payment including dates due, any interest on payment, and any discounts?
- What is the formula for sharing if a cash crop is involved?
- What is the length of the contract?
- How are amendments added?
- What happens if the property is transferred?
- Does the owner have a say in anything?
- Is the agreement binding to the heirs of both parties?
- Is it legal?
- Who are the parties, and what is the address and contact information?
- What is the exact description of the land involved?
- Does the owner have access and use of any facilities?
- Can the tenant sublease any part of the land?
- Can the tenant make improvements? If so, who covers the cost?
- Can the tenant remove trees, amend the soil, or make any other changes?
- What about introducing, or not controlling, invasive plant species?
- What is the crop production index of the land, from the government's perspective?
- What are recent soil test results?
- What is the production history (the average yield for various crops)?
- What is the history of chemicals applied to crops and herbicides on this land? (This can be a critical piece of information.)
- What drainage improvements have been made?
- Is either party flexible on percent share of production costs?
- What are the tillable acres, and nontillable but usable acres?
- What are others charging for similar land and crops?
- What is, and what should be, your return on investment?
- Are any buildings being used as part of the contract, or are they separate?
- What about tenant services (labor, fuel, mowing, repairs to fences or roads)?
- Is there good, or bad, access for large trucks?
- What are the water rights?
- What is the physical description of the parcel?
- What's next door?

Invasive Plant Species

No part of any of these programs, whether government supported or on your own, should include introducing any invasive plant species. Almost everywhere plants grow there are invasive species that are causing harm to native plants and pollinators. And almost everywhere has a list of these plants to exclude or eradicate if encountered.

But what about existing invasive plants on a piece of land you rent? Are you obligated to remove them? Is the owner? Perhaps you rented the land in the first place to take advantage of a large population of these plants—a purple loosestrife wetlands stand, for instance. The nuances should be well defined before you sign on the dotted line so you are not surprised afterward.

Purple loosestrife is an insidious invasive, and it's a profitable honey plant. What should be done with these, and who should do it? A contract should make that clear.

Making Bees

One pleasant benefit of setting up an operation like this is that with all of your bees living a fat and happy life, more bees will be the result. These bees can be used to increase your holdings, either adding to the colonies you have or moved to additional beeyards. Or they can be one more crop you harvest and sell. Perhaps this is how you pay for the land the bees are sitting on, and the honey you make is a bonus, and the crop you and the landowner harvest is an additional harvest.

We cover the basics of making bees in another chapter. Making nucs or packages, raising queens, selling brood, or even having thriving colonies will be a bonus now that you have a safe haven, and good food for your bees.

Typical leasing contracts contain at least these provisions and often more. But the focus of the contract proposed here is significantly different. This is something that's not in the comfort zone of most beekeepers.

How This Might Work

You approach a landowner wanting to rent land to grow plants that support your bees and maybe where you can keep your bees while you rent the land. The most basic plan would be: You rent the land and do all the work. How long do you rent it? For a year and you grow annuals only. You calculate costs versus yield for an annual crop or crops and see if it works. It probably will. This is kind of a cash lease deal. It's workable, but inefficient.

What about a longer-term lease? Or, an ongoing lease? Or a shared cost lease? When you open these doors the opportunities for landowner and beekeeper become almost immeasurable:

- Crops that can be harvested for cash.
- Crops that are perennials and don't need to be replanted.
- Crops that are permanent but don't take up actual crop space.
- Crops that fill niche spots and don't need any attention at all.
- Crops that government programs pay you to grow.
- Crops that provide a healthy and abundant diet for your bees.
- Crops that provide enough food, good food, and enough good food all season long.

The number of positive directions this can take is almost infinite. Landowners get soil improvement, harvestable crops, a good tenant, and income from the land and the government. And beekeepers get year-long honey crops, varietal and artisan honey crops, and a safe place to keep bees, plus they can leave the world of migratory beekeeping forever.

Now you have basic plant lists and you know where to find more lists. You know the basics of writing a lease; you know that there are government programs to help you plan this and even pay for it; you have the rudiments of planning a blooming strategy that will supply enough good food at the right time for as long as your bees need it; and now you have the safest place possible to do it all. This is, I believe, a way to keep bees that must be explored. Monoculture agriculture will continue to grow; a growing population requires more food and greater efficiency in producing that food. And that means less and less diversity for all wildlife, including pollinators.

A small, diverse, and continuously blooming settlement as suggested here is one way to combat a rapidly diminishing way of life.

Here's what you want: a beeyard safe from pesticides and close to water and an empty field, just waiting to grow you and your bees a crop of honey.

Chapter 3:

Royalty

Finding the Best Queen There Is

Keeping every colony operating at 110 percent efficiency requires many things, but most important is a well-raised, well-mated, and supremely well-cared-for queen. She must live in the best environment possible. First and foremost, though, she must be a descendent of a match-made-in-heaven set of parents and grandparents (and further back, for good measure). Mess up any one of these distinct and unrelated variables and the whole operation collapses. It is a narrow road your queen walks on, that road between success and less-than success, and the road has hundreds of intersections with other roads going to and coming from all directions. You already knew all of this, but these are the things so easy to lose sight of when time is tight, money is short, and there's lots yet to be done. Got a queen in that box? Yes, and move on. No? Darn, what to do now? It's so easy not to plan.

A major requirement, one your queen has no control over but you have complete control over, is providing the best environment possible—including both the environment inside the colony and the world outside her colony. No matter how many bees, how careful the care, or how clean the home, bees won't make honey on asphalt, field corn, or well-manicured lawns. But in this chapter, we focus on the environment inside the colony and its effect on the most important resident. We discuss the outside environment elsewhere in this book.

There she is ... she that makes or breaks your colony, and your business. If she's a good queen, you'll do well. If she's not, replace her immediately, and find out why she wasn't good.

Is the state of your equipment such that it should be out back, waiting for a fire, rather than in good repair, clean and safe for your bees and your new queen?

These are the queens we can get. The quality may be perfect, average, or they may arrive dead—a mass-market bug with no guarantees.

Royal Lineage

Unfortunately, we often take for granted the genetic background of our queens. If the ad or the queen seller on the phone says Italian, then the queen I just bought is ... Italian, right? Similarly, we often overestimate the skill, the ethics, and the colony conditions of the queen producers from whom we obtain our queens. (If she was raised at home, you could make other assumptions, though we still often overrate our own skills.)

Moreover, queen buyers (i.e., you) might make less than accurate claims regarding the skills of the beekeeper who purchased that queen (i.e., you), or the safety of the colony the queen will be introduced into. These skills include experience, patience, and the right introduction equipment. Safety reflects the presence or absence of any chemical contamination or harmful pests—and that includes everything: the wood, plastic, and wax—and the health of the bees in the box intended to care for these queens. The results of these occasional errors by the beekeeper are, however, too often assigned to the queen producer. Beekeepers screw up sometimes. But still ...

It's the most fundamental rule in beekeeping. Getting a good queen is almost never a factor of price. It is not a fact that the more money you spend the better bug you will get. Money may have more influence on raising your own queens, though: The more spent on good equipment and skilled labor to manage it the better the chances are your queens will be better.

Money isn't a guarantee for achieving excellence, just a better chance of obtaining it. Let's start by looking at purchased queens. We discuss raising your own later.

There's a rule, and it's Queen Rule #1: We settle for the queens we can get rather than the queens we want.

What Can Go Wrong with Queens?

Here is my list of some of the things that can go wrong (in a more or less timeline sequence) in queen production, sales, and installation that are avoidable but perhaps expensive to avoid. These problems begin with your queen's parents (and probably go back to her grandparents, but for now we'll stay closer to home). The problems continue to the queen you purchase, and onward through the journey she travels to get to you and eventually, to the hive.

Your queen's mom was open-mated. You have no clue who your queen's father was. However, Mom is selected as a breeder.

After the graft, Mom's larval stage was spent in a nasty, hostile environment, damaging, but not killing her. She may have been contaminated with chemicals, or fed contaminated royal jelly made by workers with access to only contaminated pollen, and not even enough of that.

Mom survives her youth, but while pupating is attacked by varroa mites in the cell.

Mom survives varroa mites and emerges, finally. She's moved to a mating nuc loaded with small hive beetle chemicals, frames with ancient drawn comb that are contaminated beyond belief with varroa control chemicals. The only consolation is that there are no small hive beetles, and no varroa.

Mom survives her few days as a maturing virgin, in spite of the fact that the mating nuc is under-staffed.

When she finally flies out to mate, she finds there are so few of the producer's drones available that she takes anything she can find. She mates with only a handful of unknown drones rather than the twenty or so she wanted. She returns both undermated and carrying the genes of unknown drones.

It turns out the drones she mated with were raised in a hostile environment of pesticides and varroa mites, limiting their ability to produce enough sperm. The sperm isn't viable, so Mom is doubly cursed: too few options to begin with, and not good enough results in the end.

Small hive beetles and small mating nucs add up to big problems when it comes to the chemicals that may lurk within. Varroa, too, cause problems in small colonies with not enough bees and the chemicals needed to control both pests accumulate rapidly.

Mom mates and her offspring (one of which will be your queen) are grafted to queen cell cups made of contaminated beeswax, which damages them slowly for the next several days.

If mom survives the rigors and near-death experience of growing up, there are additional challenges when she moves on. All manner of trials may await her before, during, or after mailing, and any of them will cause immediate or future problems; most will result in rapid, or at least eventual, supersedure.

Your already-damaged, not-yet-sent-to-you queen is then raised in a hostile environment, contaminated with pesticides, in the starter or finisher colony or mating nuc or banking colony, and is damaged beyond usefulness.

Your queen is raised by nurse bees who were previously damaged by pesticides, malnutrition, varroa mites, virus, or other maladies such that they cannot adequately care for her. She is not 100 percent when she emerges.

Foul weather prevented your queen from flying and mating during that critical window. Or …

The weather cooperates and your queen mates, but only with drones damaged from similarly contaminated colonies. Or she does not mate with enough drones, or she mates with drones that are from a diversity-poor background.

Your queen is damaged in handling when harvested (she will be superseded immediately after introduction).

While in any of these environments your queen's spermatheca is essentially destroyed by nosema, contracted in the starter, finisher, bank, or new colony: She will be superseded within ten days of her arrival.

Your spanking-new queen wasn't kept in the mating nuc/colony long enough to determine if she's mated, laying, or that her offspring are acceptable and it turns out she wasn't, isn't, or they aren't … a minimum of twenty-one days is needed with thirty-five days always better. She will be superseded in a short time.

Your queen is relegated to a holding colony—a queen bank—for days, weeks, or longer. The quality of that colony is unknown—but suspect—and all the time her health is challenged, and her sexual development declines into oblivion.

While in the queen bank the caregivers transfer any of the many viruses queens are subject to that damage, or eventually destroy, queens.

While in transit to a middleman dealer, or to you, she is exposed to heat, cold, pesticides, crushing, you name it.

Your queen is damaged while waiting to be sold from the retail dealer by being exposed to excessive heat, cold, or pesticides, or is starved because the bees in the battery pack aren't fed, or fed enough. If she is not superseded soon after introduction, it may be after she gets a round of eggs laid.

You introduce your brand-new queen into your old, contaminated, bordering-on-disgusting equipment—colonies contaminated with pesticides that the beekeeper (i.e., you) or the farmer applied. Her offspring do not fare well, the colony blames her, and she is eventually superseded.

Your new queen is introduced into your colony without proper nutritional support, or room for her to perform properly, or she manages to pick up any of the many viruses, diseases, or problems lurking within your dirty, filthy, disgusting hive. (She is superseded eventually.)

And often, your brand-new queen is introduced into a colony where there is still a queen in charge. Research suggests that as many as 20 percent of all colonies have two queens at any one time, most often a mother-daughter combination, especially in the spring. (And one of the two queens is disposed of, most often the newbie.) This mistake really adds to your cost, to pay top price for what might have been a good queen only to watch your new queen be killed because of a lack of due diligence.

After all this, it is amazing that we have queens at all, isn't it?

The Queens We Can Get

Reality check: There is no consumer magazine that publishes an annual list of "10 Best Queen Lines" or "The World's Best (and Worst) Queen Producers." Nor is there a government, and industry, or a labor group that requires continuing education credits and annual registration that monitors queen producers and their products. Moreover, there's no UL independent laboratory that measures the quality of the queens produced to see if they meet an industry standard or to see if claims made by the producer actually hold up. On the other side of the pallet there isn't an inspection service that certifies the purity of the homes queens are sent to, or the skills of the beekeeper receiving queens. So what the market accepts is what's produced and what's available. We settle for the queens we can get.

Beekeepers understand that receiving an unmated, a damaged, or an unfit queen is not the exception: It is expected. But just as likely, an inexperienced or incorrectly trained beekeeper can instantly destroy what may have been a perfectly good queen during the installation process. And in that same instant, the beekeeper will blame the queen producer for the trouble. There are at least two sides to every story, and with queens often several more.

Smaller beekeeping operations, such as those that depend on outside sources for most or all of their queens, have more exposure to queens of unknown quality than operations that raise their own queens. Regardless of the source, every beekeeper has the responsibility to provide a safe and clean environment for a queen and her colony.

Just imagine what the best queen, with the best genetics, raised in the best environment, mated with the best drones, and performing in the best environment would be capable of. That is what you need: the right queens in every colony producing bees that perform the way you want them to, when you want them to, where you want them, as long as you provide the essentials that their environment does not.

Getting good queens has never been a given. Today, purchased queens, with some exceptions, are at best a gamble, and an even bet to not be around longer than your smoker will stay lit. This is a big problem for those commercial operations that are losing 30 to 50 percent of their queens every season (sometimes more). Mobile beekeepers don't usually raise their own queens, so they use purchased queens—mated, virgins, or cells. And they continue to lose 30 to 50 percent of their queens. It doesn't have to be that way, though.

A Good Home Is Better Than Good Genes

A time-tested rule of queen keeping comes from Dr. C. L. Farrar, USDA Honey Bee Research Lab Leader in Madison, Wisconsin, who, more than sixty years ago said, essentially, that below-average queens living in a great environment (clean, well fed, well cared for) will outperform a great queen living in a poor environment (contaminated, poor nutrition, not enough population) every time.

Sixty years ago the troubles may have been different, but what he said worked then and still works today, in spite of the additional troubles we have to deal with now.

Dr. Farrar's definition of a great queen was one chosen from a great breeder, raised in a strong, healthy colony rich with young bees that had perfect flying weather for mating and had lots of drones to mate with. Get a production queen with that kind of experience for the first six weeks and you were pretty well certain to have a good colony. Sound familiar?

It turns out it's nature versus nurture. And here, nurture wins every time according to Dr. Farrar. I think he's right. If we were to update Dr. Farrar's concept to include the troubles queens face today, I think the list would look like this:

The best queen ...

- Is selected for grafting because she is derived from excellent stock (both mother and father) as a result of instrumental insemination, very controlled drone stock, or, perhaps, a wide and diverse group of drones and the genes they carry.
- Is surrounded by a chemical-free, pristine environment.
- Is cared for as she develops by many healthy and well-fed nurse bees.
- Is well-fed and cared for as a virgin after emerging from her cell in a clean environment.
- Is able to fly at will with perfect weather and predator protection.
- Is surrounded by drones in the DCA with equally exceptional backgrounds, with carefully chosen backgrounds; or is chosen to be a breeder and is artificially inseminated.
- Is able to live in a healthy environment after mating, becoming established in a colony free of diseases, pests, and contaminants.

Dr. C. L. Farrar, USDA Honey Bee Research Leader, USDA Honey Bee Research Lab, Madison, Wisconsin

What Do You Want from Your Bees?

Here's a good story—proven over thirty years of excellent record keeping by my friend Buzz who was for a long time 'bout a 100 sideliner. His comments on queens are noteworthy.

Honey production was the only metric Buzz used to grade a queen. It was the one measure that he could rely on year in and year out, and the one that kept him in business for more than thirty years. How much honey did that queen make? Of course, queens don't actually make honey, and there are a hundred things that contribute to honey production: weather, population, diseases, and overwintering, among others.

However, Buzz says, "I can pretty much handle the rest—mites, food, weather, diseases—but a dead queen doesn't do me any good at all, and an average queen is just that: average."

What Buzz wanted was honey. What do you want?

Surviving winter may be at the top of your list of things a colony should do.

Think about that for a moment. What you want from your queens may include, but by no means be limited to:

- Extraordinary honey production, being keenly aware of the many, many traits required, at the right time and in the right proportion, to accomplish that elusive metric.
- Strong pollen collection traits so your bees do a good job of pollination, but balanced with at least some nectar collection when needed.
- Propolis production geared to marketing goals or management style.
- Rapid and incredible population growth (a queen that can lay more than 2,000 eggs a day) when needed for special or seasonal purposes, such as package production or varietal honey production.

That means precisely timed buildup in anticipation of local nectar crops. Do you know what your local nectar crops are, and when they bloom (early, mid- or late-season)? Just when do you need lots of bees, and when do you need not so many bees?

It also means precisely timed mid- or late-season population reductions to coincide with those elusive local nectar flows.

And it also means they are quick to shut down during an unexpected dearth, especially midsummer (a queen that will slow to 250 or so eggs a day—primarily due to lack of food intake), and that they are able to rapidly produce a large population at an unaccustomed time to accommodate later nectar flows.

Good winter survival (not merely alive in the spring, but thriving) in cold regions requires many traits including, but not limited to, having small, but right-size populations of healthy, young bees in the fall that practice appropriately conservative food consumption; or conversely, large populations of young bees (requiring lots of food, but coming out in the spring with large populations). These traits are independent and require vastly different behaviors, but the ultimate goal is to have your colonies alive, healthy, and going like gangbusters when the dandelions (or your choice of anchor plant) bloom.

- Do they demonstrate moderate to low swarming tendencies, the ability to tolerate crowded spring conditions with appropriate food placement, and other behaviors that keep the bees home?
- Are robbing and drifting behaviors essentially absent?
- What about temperament? Are the bees fun to work, or is armor needed?
- Do they produce dry cappings or wet cappings, depending on marketing plans, or does it make any difference?
- Are they resistant to, or tolerant of, all of the pests and diseases commonly encountered? (Remember, the goal is NO chemical controls in your colonies.)

And of course, there are always more traits you can think of, or that might come in handy given your particular management style or local conditions. I trust you see where I'm going here.

"If I ordered ten queens from any supplier, anywhere, any year, and put them in similar colonies, here's what would happen almost every year: Two queens would arrive dead or would almost immediately be superseded. Six queens would produce a box, maybe two of honey,* and the two remaining queens would lead colonies that would always produce two and often four boxes of honey. That's 130 to 160 pounds (59 to 73 kg) of honey—an astonishing crop that averages about 60 pounds (27 kg) per colony per year. I've spent a career trying to figure out how to buy just those last two queens."

*For Buzz, a box is a 10-frame medium super with only nine frames holding about 30 to 35 pounds (14 to 16 kg) of honey.

Boxes of honey may be all you want out of your bees, and if you find a queen supplier that produces queens that head honey-producing colonies, then that's the best queen you can get.

Know What You Want

To get the most from your bees, you must know what you want from your bees. It's that simple and it's that difficult. There are queens that by design or luck meet some of these requirements. Carniolans are rapid spring buildup and good wintering, for instance. Italians have strong, large populations in hand, and the old adage it takes lots of bees to make lots of honey still holds true. Caucasians manufacture propolis well, and bide their time in the spring, then spring into action a lot later than their Carniolan cousins. Russians, a hybrid, variable beast, tend to be slow to build early on, then build fast, winter well because of small populations and not much food consumption, and show good resistance or perhaps tolerance to most problems. Other hybrids feature behaviors that meet some or most of these attributes. If you want a bee that stands on her head, you can probably find or make one.

Hybrids

The issue with hybrids is consistency. Breeding from them produces, often, a disappointing first generation, so you have to rely on a breeder who can successfully cross and cross again to get what you want. A while back the Buckfast bee was like this. If well done, it was a great bee ... if poorly done, it wasn't bad, but it wasn't Buckfast. If your goal is to consistently produce a consistently performing hybrid, then backcrossing and all manner of tricks can be employed that might work. Or you can get a degree in honey bee genetics and still hope they mate with the right drones. Otherwise, it's a process of flooding your area with all the genes you want so no matter where your carefully chosen virgin queens mate, they mate with your carefully chosen drones.

Survivor Queens

And there are those "locally produced" queens, sometimes referred to as "survivor queens" because they seem not to be bothered by problems. They are usually mated with "locally" produced drones that produce "locally" adapted bees. It's a pretty straightforward process that doesn't require a lot of training: Raise your own, let them mate, keep your drone colonies well populated, give away queens to local beekeepers, and choose the best of what comes. You'll get good, mostly resistant, pretty productive queens, most of the time. And it will take some time, so be patient.

So I'll ask again. What do you want from your bees?

Emergency Queens

Many beekeepers will tell you that supersedure queens are not only acceptable, but sought after because they are, almost always, raised at the best time of year by bees who know they need a new queen. Moreover, their genetics are certainly acceptable, and their care and feeding during swarm season is usually more than adequate. Add to that the fact that the drones they mate with are the drones you want your queens to mate with, and the resulting queen should be just fine.

If all of those conditions are in place, I agree. But if any of them are missing, then you have lost some degree of control, and it is this control I think you should be striving for. Some colonies will swarm, no matter what you do, and some colonies will not be satisfied with the queens you provide and want their own. It happens. It's best if you know it happens though, so when evaluating a queen, you know if you are evaluating the queen you put in that box. A marked queen makes that task easy.

Emergency queens, however, are another story, and should not be tolerated. You have no idea of the age of the larva selected, the diet she was raised on, or with whom and where she mated. You have given up any chance of control. Replace emergency queens whenever you find them.

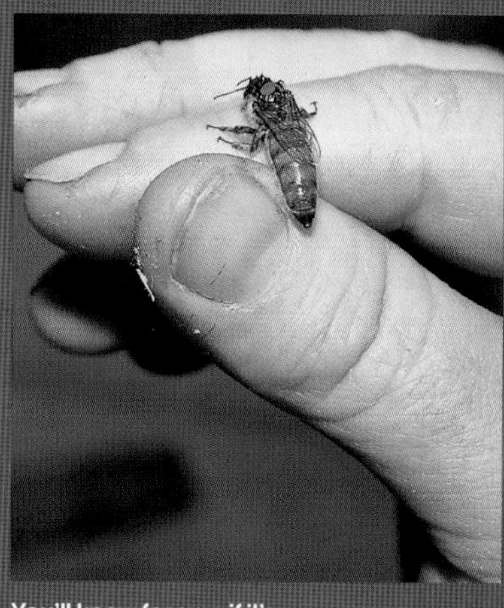

You'll know for sure if it's your queen, or their queen, if she's marked.

What Kind of Beekeeper Are You?

Like Buzz, some beekeepers keep bees just for the honey. But by now you know what it is that you want from your bees, whether it's money, honey, wax, easy-to-manage populations, pollination, gentle temperaments, don't have to feed, don't have to medicate, they stay alive ... the list goes on. If you haven't figured it out, now's the time to do so because without that critical bit of knowledge, you might as well be tending sheep.

But what kind of beekeeper are you? If you are out of town on business every spring and your bees swarm every spring, those aren't the right bees for you. What if you want your honey sales pitch to be "natural," but you have to keep treating for pests and diseases to keep them alive? Maybe you can't stand scraping propolis, or really don't like getting stung, or can't afford having to feed every winter?

You and your bees have to be well matched and function as a team or you end up fighting your bees, or trying to keep up with your bees, or trying to get your bees to keep up with you all the time. In short, trying to get them to do things the way you want them to when they are programmed to do things another way is a sure road to disaster. So be honest.

Two-queen colony

Backyard beekeeper

Serious overwintering

Running the store is okay, but I'd rather be working with bees.

I keep perfect records.

Sometimes the weeds get ahead of me.

What Kind of Beekeeper Are You, Really?

Evaluate your lifestyle, your personality, and your habits (good and bad) and choose the qualities that fit your beekeeping style.

- Are you slow to start in the morning, in the spring, at harvest, and for winter prep?
- Are you a fast starter in the spring, but get bogged down mid-season with farming, job, gardening, vacations, kids?
- Are you fast in the spring and summer and fall (always fast, always ahead of the curve, start early, done early, looking for more)?
- Are you good at fixing things, making things, scrounging things, finding things?
- Are you the type of person who couldn't figure out a screwdriver if one end wasn't pointed?
- Could you sell ice at the North Pole?
- Are you the type of person who couldn't sell ice at the equator?
- Do you really enjoy the time with the bees, being outside, sunshine, rain, moonlit moves, building or fixing equipment, being with beekeepers, the smell of wax, sticky floors, and even beekeeper meetings?
- Are you a fantastic record keeper, have a great business history, have a good plan, and can stick to it?
- "What records? What plan?"
- Do you want to make all the money you can and spend as little as possible, cheating the bees, your family, and the government to do it?
- Do you have the attitude of, "Life is too short … slow down, enjoy the day, the bees, and each other"?
- Are you all about business (contracts, deals, buying bulk, spreadsheets, bottom line, buy low and sell high, aim for the least expensive means that meets every one of your goals)?
- Do you need bees really early to make lots and lots of honey, queens, nucs, and splits?
- Do you like tucking the bees away for five months and taking the winter off, spending it where it's warm?
- Are you up for moving bees two or three times a year (it is good for you, good for the bottom line, and there are no family or other anchors to get in the way)?
- Are you thinking, "Honey crops are long and slow and late, but so am I"?
- Do you need lots of help with the bees from family, school kids, friends?
- Do you work alone and like it like that?
- Do you forget where half your beeyards are half the time, so you want bigger and fewer yards?
- Can you sometimes squeeze enough time out of your day job to get all the work done on time, and all right? And sometimes you can't?
- Are you looking to get every last drop of income possible out of every box, everything but the buzz, because you have kids to feed, a house to pay for, and no retirement account to fall back on in five years?

Be sure to get queens that conform to you and your operation, your location, and your style. Don't get queens that conform to the operation you wish you had or thought you had, or might have had if it had worked out last season, or last winter, or if your other job hadn't fallen apart. Be realistic. As one beekeeper told me years ago, "If you sleep in, make sure your bees do, too."

But where do those queens come from?

Buying Queens

As you already know, there's more than one way to get queens—some better than others.

Mated Production Queens

Well, you are already likely buying mated production run queens from queen producers. You've probably been doing this for years. It's the fastest, most expedient, easiest, most common, and often both the most disappointing and the most promising way to obtain queens for your colonies. You probably have a favorite producer or two. And you probably order the same kind of queens each time from each of your several suppliers. You seem to have a good enough mix that way.

Those old standbys—the three banded Italians, Italians, hybrid Italians, mite-resistant Italians, Cordovans, VSH, All American, Buckfast, New World, hygienic behavior, Russians, hybrid Russians, or Carniolans, fresh queens, or just queens—are advertised in every journal all year long. Not much information is provided. You take what you can get and like it.

But recall Buzz's bell curve experience, the list of needed attributes I provided, and the years and years of your own experience. Given that there are many attributes to choose from, and every year there are fewer suppliers (how many of the suppliers you bought from ten years ago are still in business?), what do you really think about buying already mated queens? What do you think about how much you have to pay for them? The variability of the queen stock in the United States, for example, is decidedly limited. Far too few breeders are supplying breeder queens to an ever-decreasing number of production queen producers. The bee industry is shooting itself in the foot. Programs all over the world are doing the same thing because varroa has killed so many bees, both feral and managed, that only highly managed, highly protected bees are left.

But, there is an upside.

Though buying mated and unknown queens can be expensive, just getting any queen you can get may be an acceptable way to "hold" a colony until you can obtain the queen you actually want to install in that colony that produces bees with the specific attributes you are looking for. (Remember, this assumes that now you have decided what you want from your queens, and what kind of beekeeper you are.)

If you have lots of colonies, and you have to buy them at retail price, the queen cost on the ledger will be high. Still, that total will be

less than losing colonies that go queenless. Crunch the numbers and see if this approach makes sense. It's marginally effective from a financial perspective, and probably not effective from a quality perspective—even if the placeholder queens are there for only a short time. Don't forget, though—now you are doing this for far more colonies than last year, and have less time to do it in. That absolutely must be considered: It's that old time or money thing again.

But keep in mind all the troubles inherent in the above system. Queens that are produced by others are subject to the chemical, environmental, and ethical problems already mentioned. (Perhaps Buzz's bell curve is related to this?) And, once you get the queen you want, you can still use the other queen you could get for other purposes, like holding another colony until you can get the queen you want for that colony. Queens from others are barely worth the effort, but an option to keep open.

Uncapped Queen Cells

Yes, you read that right. When you buy uncapped queen cells, you are essentially buying the genetics. The cell producer has very little invested in this animal: a graft, two or three days in a strong starter colony, and shipping costs. That little investment means there's been very little time for things to go wrong on the producer's end, and almost all of the responsibility rests with you when it comes to raising these cells. And they are cheap.

Those who buy uncapped queen cells routinely report that final acceptance of mated queens is a bit lower than the percent of capped cells, but only just a little. And the cost is far less, so buying more than you need—finisher colonies will accommodate some loss (about 20 percent)—is the way to go. Besides, then you can select the best of the lot and let the small ones go.

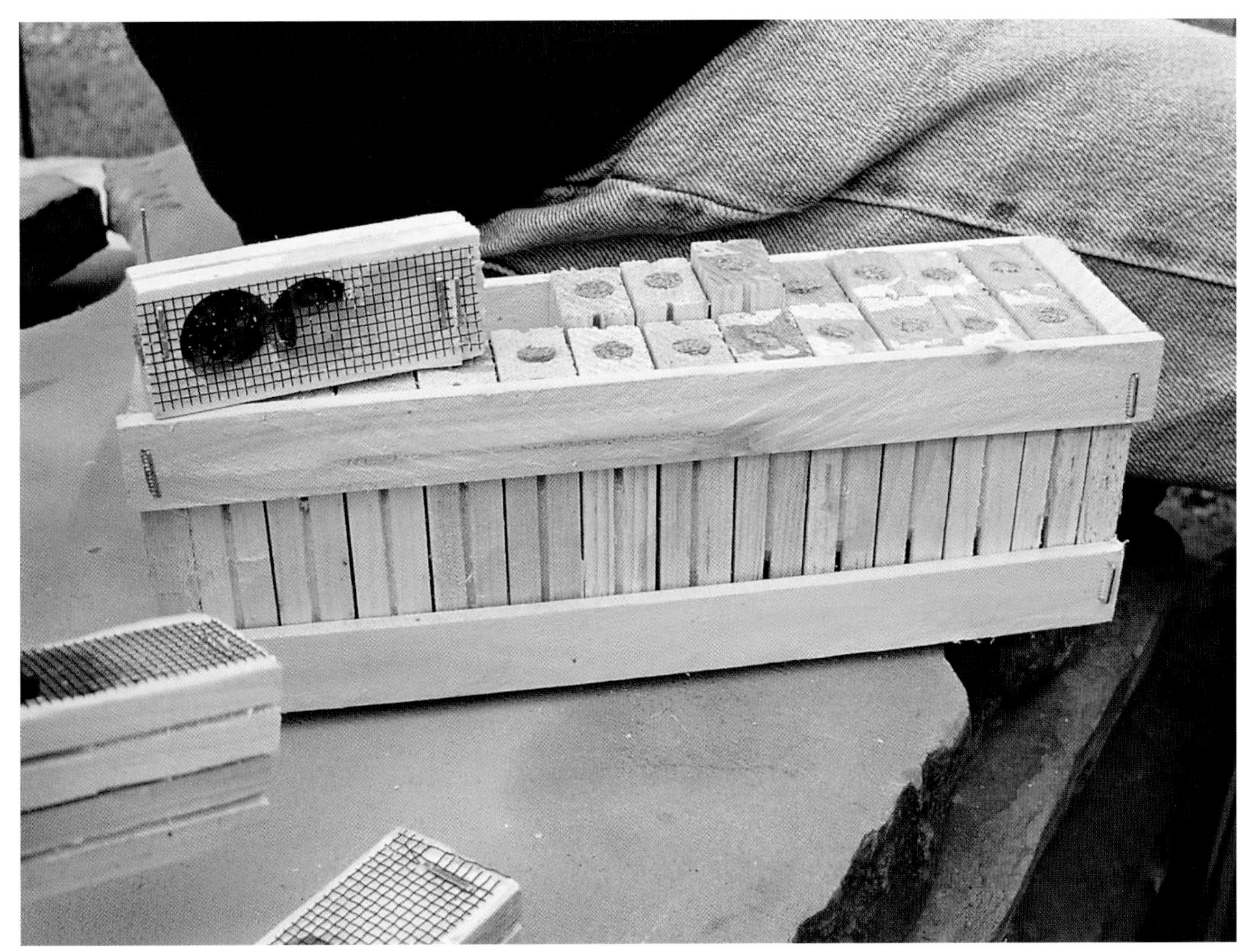

Is this an Italian because the supplier says it is, because it looks like one, or because it behaves like one? You'll know in a few weeks, when it's way too late to do anything about it.

Getting Queen Cells

Queen cells are inexpensive, generally easier to obtain at the right time of year and will have mostly the right genetics if you know your producer. When these queens emerge, they mate with the drones you provide.

Like open queen cells, some of these queens won't make it to royalty but that number is usually small. Cell producers have environmental issues that can damage a queen's development, so some won't emerge. Some emerge and, though initially accepted by the colony as virgins, they do not mate or are lost during mating. Cells placed in queenright colonies are usually doomed, and poor flying weather or becoming a mid-day lunch for a hungry bird may become problematic too.

If the colony was queenless when the cell was introduced (and it should have been) losing the cell (or the virgin) usually results in a hopelessly queenless colony that quickly needs another open cell, cell, virgin or mated queen, or to be combined with a queenright colony. The overall average of lost queens from cells is between 15 and 20 percent—higher than most beekeepers are aware of and producers will admit. If that lost number is less, the cells were most likely from a good producer and you did a good job of introduction. But if the loss is more than that, and that happens just as often,

Buying capped queen cells gives you a bit more control on the outcome of the colony this queen will lead because at least you supplied the drones she mated with, and, importantly, she was raised in your clean and safe environment.

there may be something wrong with the cells, the cell producer, or, just as likely it is a beekeeper problem.

Low Initial Cost

The low initial cost of purchasing queen cells rather than mated queens or raising queens yourself has to be figured into the equation. You can probably live with that loss factor because both the availability of more cells and the total cost are acceptable. Queen cells can be replaced quickly—either with another cell or even a mated queen (a holding queen)—and the cost of cells, or even a much smaller number of mated queens, is low. Consider, though, the time needed to get the colony up to speed using each of these sources. A mated queen is the shortest and if the honey flow is only two or three weeks away, losing that colony's production—let's say it should have made 40 pounds (18 kg) of honey on that flow—needs to be figured into the cost of replacement. The open cell is certainly cheaper—but not producing honey is expensive.

Genetic Control

A decided advantage of purchasing cells is that you have some control over the genetics of the requeened colonies. She is, after all, mating with your carefully selected drones. If there are good genetics and a healthy environment on the producer's end, and good drone genetics and a healthy environment for her to live in after mating, then purchasing queen cells should be on your agenda. It poses less risk than using open cells, and is a cost-effective way to not raise your own queens.

Acquiring Virgin Queens

There's a classic technique that is resurfacing that bears investigation: buying virgin queens. It's not a lot different than buying queen cells, but it's slightly less risky. The virgin queen producer has more time invested in her, so the cost will be more, but the possibility of handling or shipping damage is reduced. Introduction is the same as for a mated queen: Introduce the virgin into a queenless colony in a protective cage, and keep her there for four to seven days (closer to seven for my taste, but often four is sufficient). Only release her if there is no balling behavior, and from there on biology takes over. She will take anywhere from a day to five days or so to start mating flights, and will, weather permitting, finish that in three or four days and then set up shop. You should have a laying queen in ten days or so, and a good laying queen in two weeks from arrival.

You have more control of mating with a virgin you can release at will than with a loose queen from an open cell or queen cell because you can release her when the future weather looks favorable, rather than not. And of course, your control of the drones she will encounter (see Raising Drones, pages 93, 98, and 112.) remains strong. Plus, you can mark her and know for sure she's the queen you bought.

Raising Your Own Queens

Raising your own gives you more control, but at the same time more responsibility, and there's nobody to blame for problems that arise. But this is the only road to success.

Getting Breeder Queens

When you choose to raise your own queens you have to start somewhere. Here's where.

Breeder Queens from Somewhere Else

If you are going to raise your own queens, you first need queens to produce the queens you will raise. One choice is to purchase already-mated breeder queens from actual queen breeding operations. These very expensive bugs often start in a small colony or nuc to ensure acceptance. Once laying and seemingly happy, the nuc is used to requeen a small colony, which is then boosted with lots of brood and young bees from other colonies to ensure ample support for the new queen and her care.

Once your breeder queen is laying an average of 1,000 or more eggs per day, begin using her to make production queens by any of the techniques discussed later. There are easily a hundred ways to raise queens. I suggest you try as many as possible until you find the exact technique that you like, works best, and fits your operation.

The production queens you raise will mate with your drones in your area and eventually produce the diversity of bees you want. You may have breeder queens from more than one source along with drones from other sources to produce a diverse and varied population within each colony, and within your operation. Or, you may choose a more defined breeding population for specific purposes.

Breeder queens must be produced in a friendlier and safer environment and undergo more rigorous and longer evaluations than run-of-the-mill, quick-out-the-door production queens. A higher standard is to be expected. Well-raised, well mated, well cared for, and complete with an excellent genetic background should always be part of the deal. These production queens then are raised in your environmentally friendly starter/finisher colonies and finally introduced into your safe and sane colonies. That gets you the right bees.

Breeder Queens from Home

You can, and probably should, be selecting at least some breeder queens from your own stock. These are chosen by any criteria you select and will produce virgin production queens by natural or grafting methods that will mate with drones of your choosing in your area that eventually make the bees you want. We'll briefly review a couple of popular methods: the standard starter/finisher technique and the effective but less well-known

As Mark Twain might have said, "The difference between queens you buy and queens you raise yourself is almost always the difference between light and lightning."

How many capped cells on this frame? Count fast, and then count all the rest in the colony, and you'll know how many eggs the queen has laid in the past 12 days. Divide your count by 12, and you'll know how many eggs per day—is it more or less than the 2,000 you always hear?

Cloake board method. The methods are different, but the biology is the same. These are not the least-expensive methods to raise queens. But I have watched lots of operations and these are the methods that seem to work the best for a variety of reasons, primarily because they work with the bees' biology.

There are hundreds of books on how to raise queens using almost as many techniques as there are books. You probably own five or six, and have read more than that, plus watched dozens of instructors walk you through the process. What separates the most successful queen producers from those of mediocre quality is quite simple—being able to count, and having colonies with more bees than you can imagine needing. Basic queen-rearing biology says it takes so many days to raise a queen, no more … no less. Miscount and you've screwed up the system. And having enough bees of the right age, in the right place at the right time, is about as straightforward as can be. Not, perhaps as easy as can be, however.

Drones and Genetic Diversity

There is one other variable you need to consider: drones. Most books say to make sure the drones that your carefully chosen queens will mate with are equally well chosen. After all, the drones provide the diversity a colony needs to hedge its bets against the many problems it will run into. Without a good selection of genetic material, all the workers would be essentially the same, which offers no genetic diversity. For a small operation, there simply aren't enough colonies to spread around to ensure queens from your mating yard will mate with drones you want them to meet. Or you have to spread them so thin that you spend all your time in travel and not in the bees. And you still don't have enough drones in the right places at the right time. But with your expansion this should become less of an issue. Still, too often the trouble with queens is caused by troubles with drones. You absolutely need enough healthy drones in the area.

Without healthy drones, the best virgin queen in the world is no better than a lazy worker.

Calculating Brood and Eggs

If you haven't calculated your queen's rate of brood production before, now's a good time to learn. It will come in handy in a lot of ways in your management plans, and be a tremendous help when evaluating your queens or calculating colony growth metrics.

After your new queen has been installed and accepted in her final colony, give her two weeks to acclimate and get in the groove.

After that, count the amount of sealed brood in the colony. All of it. Recall that a deep frame has roughly 4,500 cells on one side, give or take (count your frames, don't take my word for it), so when measuring, estimate the percent coverage of sealed brood each side of each frame has. It shouldn't be hard, but it takes some practice.

Here's a trick. Go through a colony with a helper. Both parties do the estimating. For instance, when looking at a frame with a typical brood pattern of a center half circle, estimate the percent coverage, and both write it down. Then, take a photo of each side of each frame counted. Label each side of each frame with a sticky note so it's in the photo, and when taking notes use that number to record your estimate. Digital cameras make it easy to photograph and refer to later. Keep your original estimates, then count the cells later, using the photographs. After only three or four frames your skills at estimating become quite accurate. It's a valuable skill.

Once you have this first count, wait 12 days, then count all the sealed brood again. This second number is the total number of eggs your queen has laid since your first count. Divide this total by 12. The result is the average number of eggs per day your queen is laying.

If you want to know if her rate is increasing or decreasing, do this three times. Each time, subtract the last brood count from the present count, then divide by 12. The result will be the increase, or decrease, in eggs laid per day since the previous count.

Just remember to do this in 12-day increments. It's a good management practice and the numbers will tell you a lot about what's going on both inside and outside the colony. And it gives you yet another metric with which to evaluate your queens.

Raising Production Queens

Here are some problems you'll encounter.

Random Queen Production

Have you been producing queens by finding swarm cells, supersedure cells, or jeez-there's-a-cell cells; harvesting them from wherever you find them; and putting them in colonies, splits, or swarms that need cells or queens? If so, stop now!

This haphazard and reckless method of choosing a leader is unsafe at any speed. If a queen cell isn't produced by you, it's an accident. Queens and the drones they mate with must be the result of your efforts, your work, and your choosing. A swarm cell is the result of your mismanagement. A supersedure cell is the result of your mismanagement. A queen cell for no reason is the result of no management. The queens produced from any of those cells are virtual unknowns from an age-related and genetic standpoint. But worse, the selection and the care and feeding they've received as larvae is totally unknown. Here's why.

You don't know the age of the larvae that were chosen. Some may have been only a single day old but some may have been as many as three days old. The three-day larvae that were selected will emerge first because they have the shortest maturation time, and they will without a doubt make the worst queens.

First Is Worst

For unplanned cells simply remember, first is worst—because they will kill the others. The (perhaps) better choices didn't have a chance. If you've been doing things right you may have a feel for the health of the workers who had to feed her. They may be clean, healthy, well fed, not been exposed to pesticides at home or abroad, and free from stress and abuse. If you're provid-

The only way you will know if the queen you have is the queen you are supposed to have is if she is marked. If not, it's a queen you shouldn't have, who shouldn't be there, and should be removed and replaced.

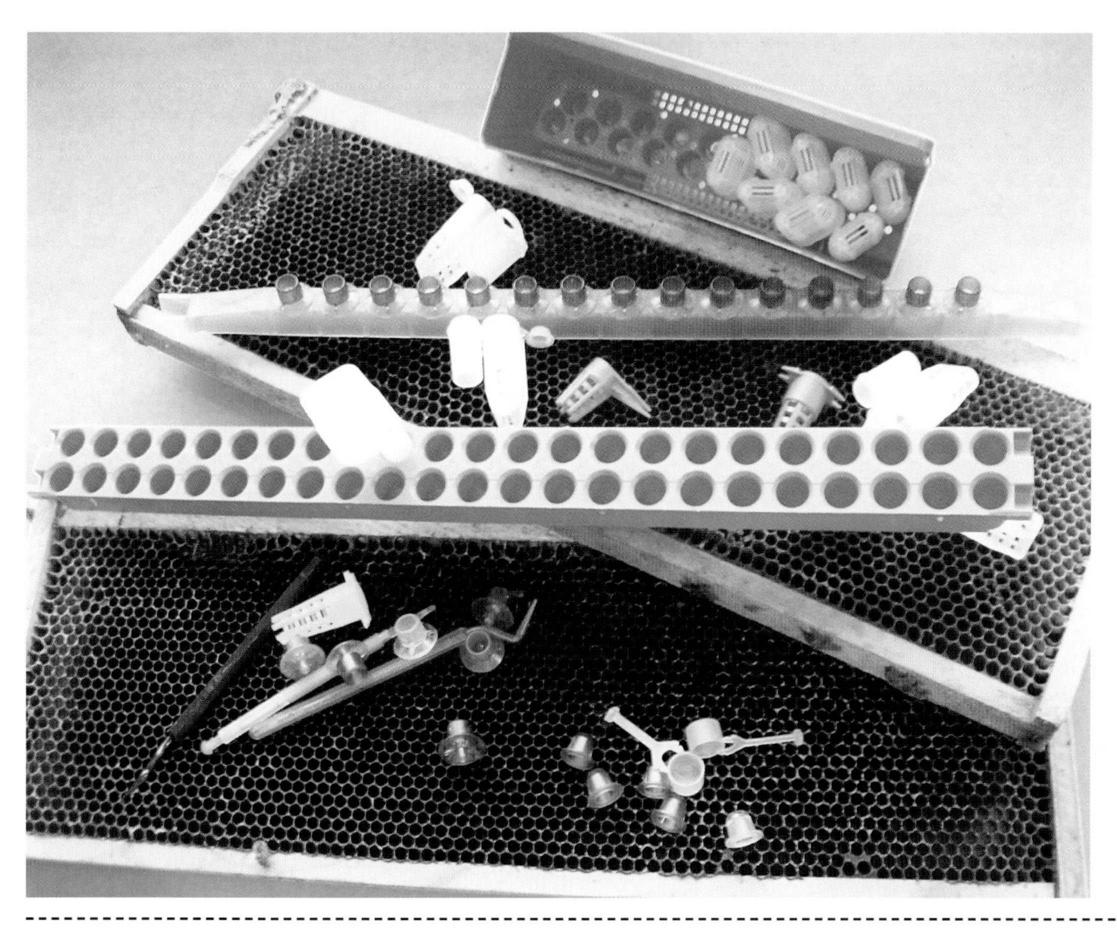

There is a lot of specialized queen-rearing equipment you can buy. This plastic equipment goes a long way in making queen production easier and safer. The golden rule for this task is to avoid using contaminated wax cell cups.

ing all the forage in a safe location you can feel a little better about this. But if you're not, you have to look at the food reserves available between four to eight weeks ago to know if the workers were well fed, healthy, and able to produce the best royal jelly. Do you know how much pollen was available six weeks ago, and what its quality was? What about nectar? Or what the weather was back then?

Plain and simple: If you don't control the way your queens are raised from eggs to egg layers, those queens are out of control, and frankly, you do not know what went into their production. They're worse than a crapshoot; they're just crap. And using them is gambling on the future of your business. Is that a business plan your banker would approve?

How much control do you have when buying mated queens? None. If you are buying queen cells you have some control, but only over half the factors at most. If you are raising your own queens, you control everything:

- Grandparents
- Parents
- Mother
- The eggs the mother lays
- Quality of the diet your queen larvae receive
- Health of the care-tenders
- Cleanliness of the environment in the hive
- Number, health, and genetics of the drones the queen will mate with

If you raise your own queens properly, you can have it all, and the future of your business and the future of that colony will go hand in hand because you have control. Give up that control and you lose. Keep it at home and you win. It's that simple, and it's that complicated.

The Russians Are Coming

I like pure Russian bees. They think like I do in lots of ways. For instance, I have a cistern at my house that supplies us with all of our water. When it rains, water comes off the roof, down the eaves, and into the prefilter tank. From there it is filtered and goes into the holding tank and is pumped from there into my system, into both the bathroom and the kitchen. I'm frugal with water.

My Russians are frugal with their resources. But when times are flush, so are they. (I'm still frugal with water, however, even in wet years.) When there's food, the Russians feed the queen, and a typical Russian queen can lay like mad. Populations will seemingly explode. I have not done an eggs-per-day count on my current queens, but a couple of years ago one of my pure Russians was averaging well over 1,600 eggs a day. She went down to fewer than 400 a day when the summer dearth came later. Those queens can really kick.

They have another trait—sometimes, in some lines, some years—that is disconcerting and takes some getting used to. Apparently, back in the day these bees were subject to springs that were tough, erratic and difficult to survive. Losing queens must have been a typical response to the elements, because now, some Russians build queen cells, especially in the spring but for some, all year long, like there will simply be no tomorrow. I have found more than twenty cells in a five-frame split in mid-spring. I've seen ten or fifteen in strong colonies, making honey with lots of room and a six-month queen—in midsummer. They take no chances on going queenless.

One of my first, and one of the best pure Russian queens I bought. I love the way she looks, and temperament of the bees was superb. Russian hybrids, at least first-generation hybrids, are usually an altogether different animal, however. If buying, avoid them. If breeding, keep the colony the hybrid is in small, and simply use her for a breeder.

Some Mated Queens Are Good Queens (Here's Why)

I'm pretty down on buying mated queens because mated queens are mostly a crapshoot. But, not completely. Some queen producers do a good job of producing mated queens for a reasonable price. You probably know one or two that you are comfortable with because the queens you get from them produce colonies that are consistently productive and healthy (but still with that bell curve of production that Buzz looked at, I'll bet).

It's tough raising perfect queens all the time. And mostly, it's the weather that gets even the best of them. It's the weather that will get the best of you when you start this activity in earnest, believe me. You can control the genetics, the health of the caregivers, the cleanliness of what's inside the colony, reduce to almost zero the chances of being exposed to the harsh realities of what's out in the real world—but you cannot control ten days of rain. Neither can queen producers. Period.

But here's the difference between good and don't-buy-from producers: The good producers leave their queens in mating nucs long enough so that they can see the first flush of brood. That takes time—and it's time or it's money. If you want a queen that has been tested to see if she is producing what she should be, then it takes longer and costs more. Are two lower-priced bugs that crap out cheaper than the pricier bug you raised that doesn't?

If you have a queen producer that you trust, ask about the weather when you are buying their queens. Better yet, watch the weather in the region where they raise their queens when the queen you are buying should be out mating. Then buy only tested queens. If the weather has been quirky, a tested queen will always tell. A quirky queen will always be asked to leave.

This was one of the good ones. She was advertised as an Italian, looked like one, behaved like one, and produced like one. She was everything I expected, and I got what I paid for from the commercial queen producer I bought her from.

Raising Drones

Drones are without a doubt the weakest link in the chain of producing perfect queens.

The only time you know for sure the background of the drones your queens are mating with is when you have carefully selected their mother, tested her other offspring for the traits you are interested in, checked the consistency of her offspring, and then made a single-drone instrumental insemination mating. Then you know. Otherwise you don't. But that's unrealistic when running a busy business.

You can stack the deck in your favor by flooding the area your virgin queens inhabit during their mating flights with drones you are comfortable with. This way your queens get an acceptable level of appropriate couplings. How many appropriate matings do you need to get a queen that produces the bees you want?

It comes down to the drones. To get the right number of drones, of the right age, in the right place at the right time in the right condition is probably the most complicated part of this task, and it deserves some attention.

The USDA determined that to maintain European colonies in areas considered occupied by Africanized honey bees, a beekeeper should have 60 drone mother colonies in the surrounding area for every 1,000 mating virgin queens. This same ratio would be required in any area if you want your virgin queens to mate with known drones you have bred and placed in the drone mother colonies in the area.

This saturation level is beyond the reach of nearly every backyard and small-scale sideline beekeeper. Think of the resources this would require in just the drone mother colonies. You can fudge those numbers if you are not raising 1,000 virgin queens, but you can't fudge them much because this many colonies not only provide the number of drones needed (20,000 for 1,000 queens), but they also cover the entire area your virgin queens will be mating in. So not only do you need density, but you also need area. It isn't as simple as ratcheting down the number of drone mother colonies needed to accommodate your 100 queens from 60 to 6. You still should be covering the same area those 60 colonies would cover.

Here's something that is seldom mentioned. If you have, say, five breeder queens producing production queens, you should have some of those same genetics represented in the drone population you present to those virgin queens. But if there are too many, inbreeding creeps in. So, what drone genes do you supply? Here's a way to think of this. In your evaluations of many breeders you will test, say, fifteen breeders from several sources. Of these, six or seven will make the grade. Half then go to producing virgin queens and half go to producing drone colonies. That way all the good traits are used, and inbreeding is reduced to nearly zero.

Insurance Policy

You will need insurance. No, not a business policy to pay premiums: The insurance you need is the extra drone mother colonies in the area that will ensure your virgin queens are mating with your drones. The competition for virgins is tough in the drone world. If you've seen a drone comet in a DCA you know there are 30, maybe as many as 200 drones vying for a chance at the dance of death each time she flies by. You want the majority of those to be the gametes of choice chosen by you, which is why you have

Find the drone. Drones are hard to find sometimes, and if you don't have enough of them when needed, even the best weather in the world won't get you a good mating. You have to work to have enough drones—the right bees, at the right time, at the right age.

For new beekeepers those packages from afar can cause an issue with alien genes. But overwintered nucs are becoming more attractive to both sellers and buyers in many areas.

drone mother colonies in the first place. But the number of feral colonies is growing because of increasing varroa resistance in the feral population. And, there are other beekeepers in the area, probably more than you know about. So how many of these unknown gamete contributors are out there?

You can't do much about the feral colonies right away. But you can do something about the managed colonies (and eventually many of those feral colonies, too). What can you do? Give every beekeeper you can find all the queens they want and need. For free. You need essentially the same resources to raise 10 queens as 25, and to raise 100 as 200. Why not provide everybody with the queens you want, especially established beekeepers close by?

For new beekeepers those packages from afar can cause an issue with alien genes. But overwintered nucs are becoming more attractive to both sellers and buyers in many areas, so there are more of them, and more people willing to buy them. If you are selling nucs locally, with your queens, at an attractive price (think cost and eventual contribution to your bottom line, not profit) your investment in both giving and selling local queens pays dividends every day. These dividends are paid not only to you but also to every beekeeper who will have bees bred specifically for his neighborhood. And where do feral colonies come from? Swarms. What colonies are swarming? They're actually your colonies, if you think about it. You've simply added to the number of drone mother colonies in the area. There are costs to this, but the result is a benefit to your breeding program, to the quality of the queens you produce, to the quality of the bees growing in your area, and to the benefit of every beekeeper in the area.

Since the donation of these queens (and thus their drones) directly affects your ability to produce good queens, the cost of producing them and the queen herself is a business expense—with a positive tax advantage. You are producing these queens for a definite and business-related purpose, so there is no reason not to include the cost of that in your business model.

Selecting for Traits: Where to Start

No matter where your queens are coming from—your operation, a friend's outfit, a reputable producer of production queens, a cell producer, a breeder producer—get rid of the assumption these breeders will produce good queens. You must test, measure, and evaluate the results of the matings your queens make.

First, decide on the traits you are looking for and how aggressively you will pursue them with the queens you produce. There are reputable breeders that produce breeder queens with different traits and attributes. Investigate and see what it is that your operation will gain by using their stock. What are their criteria for selecting from the many desirable traits? Their stock may be just the thing for you if it is well chosen and maintained, and it may save you two or five seasons of selection work; when you look at the cost of a very expensive, instrumentally inseminated queen to add traits to your line, that investment becomes quite insignificant.

The USDA program in Baton Rouge, Louisiana, has a good primer on selection criteria, as does the Minnesota Hygienic Program from the University of Minnesota, along with the New World Carniolan Project. There are several businesses that are equally well run but not associated with a government or university program.

Many beekeeping associations around the world are running selection programs for their areas, including several state groups in the United States, Europe, and South America. Some of these are working to restore the traits of bees indigenous to their area, such as the British Black Bee Project, while others are working to produce hybrid bees that will perform well in the local climate and geography. All of these operations make choices, evaluate the results of those choices, and continue to improve their stock. These operations advertise in their respective journals and are known throughout their industry. Generally, they produce a consistently high-quality product year in and year out.

There are a multitude of traits you can select for when working to produce those perfect bees. The biggest problem we have is varroa mites. Bees that are tolerant, or resistant, to varroa will live longer, suffer from fewer problems, make more honey, cost less to manage, and be more fun to have. This is one of the traits you should be looking for. This frame shows the results of a severe infestation. Note the abundance of white fecal spots near the mouth of the cells. This colony died because of the damage varroa did to the bees that lived here.

Finding the Right Bees

There are nearly as many ways to find the right bee as there are breeders, but they more or less boil down to making choices in two areas. The more traditional method is to begin with a large number of colonies, headed by queens from a variety of backgrounds. The colonies are left to their own devices under the management scheme favored by the breeder and location and at some point in time all (remaining) colonies are evaluated for the trait(s) in question. Many of these programs begin with the most important criteria: selecting for survival from varroa.

The main technique is to not treat: no essential oils, no drone brood trapping, and certainly nothing resembling a hard pesticide. Generally, a great many colonies die when they are not treated, some sooner and some later.

Those that survive, then, are the core of the selection group, and it is from these that the remaining traits are found: gentleness, overwintering, honey production, and the rest. It is an expensive and difficult means for finding what you are looking for.

There are other techniques. In one, many lines are worked while simultaneously establishing many of the traits you are interested in: gentleness and overwintering, as well as food consumption, swarming, hygienic behavior (this may, or may not, be a good choice for your operation) that controls most diseases, spring buildup speed, fall collapse timing, resource utilization, and so on. When all of the chosen traits are predominate in the still-mixed but by now reduced population, the final, usually more lethal choices, are looked at: resistance (or tolerance) to varroa or other diseases or pests. This way, all surviving bees already have the favorable attributes you want and far fewer bees die needlessly when looking for varroa survival—the slaughter of the first borders on criminal.

When selecting queens to become breeders, you must evaluate the entire colony, every time you visit. Temperament, any signs of diseases or pests, the brood pattern of the queen, and the rate of egg laying over the course of the season are all important. Honey production certainly is on the list for early, mid, and late honey crops, and overall total—and you must compare your figures to other lines, and to the standard (the line you've been using because it was the best). Evaluating colonies should be on your to-do list all season long, and the costs built into your overhead. No excuses.

Thinking Locally

The queens you produce should reflect the needs of your program. Those queens will come from good stock, the result of carefully chosen queens and drones mated by instrumental insemination, from good free-mated commercial stock available from breeders and commercial queen producers you know, and from reliable sock selected in the field by you and beekeepers you know and trust. All this stock must be tested for all the traits you are interested in, in your location. Stock that is advertised to winter well—in the far north, tropics, desert, country, city, drought region, rain forest, or anywhere—may not be the best for where you are. You know what local conditions your bees must thrive in, and the kind of beekeeper you are, and the beekeeping you do that your bees must also thrive in, which means you have to measure. You have to find exactly the right bees for your management, location, and business.

The trait that may be the most difficult to cultivate is longevity. In the past three decades, queens that survive two years is a miracle, but fifty years ago, queens that lived to be five years old weren't uncommon. Breeders haven't forgotten about this trait, and while shopping for genes, stay on the lookout for this. Ask how long their queens live. If they don't have a minimum of three years as a productive queen, call someone else.

Once you have brought the right combination of genes into your operation the evaluation process begins … and frankly, it never ends. The world continues to change, so the bees you use and the way you manage them must change with it. Some degree of time and expense must constantly be employed in evaluating your colonies. Evaluation costs should be as much of your overhead budget as fuel or leasing agreements.

So, like Buzz, if it's simply honey production you are interested in or if it's much more complicated, involving pest and disease resistance, gentleness, longevity, wintering ability, or whatever is important to your operation, the measuring never stops. All beekeeping is local, and what you choose should fit you like a glove.

The goal is zero tolerance for those traits you don't want, and 100 percent for those you do. Ask the good producers what their criteria are and how close they come. You will be surprised at how good those queens are for the traits they select for. And every year, have a test yard. Test. Measure. Evaluate. Discard.

The Starter/Finisher: The King of Queen-Rearing Techniques

The biology of the starter/finisher technique is straightforward, and the techniques were developed by Doolittle decades ago so it is often called the Doolittle method. The fundamentals are simple and most beekeepers already have the equipment they need. Some version of the starter/finisher technique is practiced by the majority of queen production operations around the world.

Note: All starter and finisher colonies use wax previously obtained that has not been contaminated by in-hive pesticides for mite control. These chemicals include coumaphos, fluvalinate, and the metabolites of amitraz. There will ultimately be environmental contributions to your wax in the starter/finishers no matter how careful you are, or how much you dictate the foraging activities of your bees. But this fact cannot be overstressed: Use the cleanest wax you can obtain for these colonies.

A whole yard of finisher colonies and support colonies for breeders and young bees

When cells are finished, they are often moved to small mating nucs at a mating yard.

Starter/Finisher Basics

■ To make certain the population of bees in an operation remains diverse and avoids inbreeding, you need to use and continue using a diverse selection of breeder queens, some from your own operation that have shown their mettle under fire in the field, some from commercial breeders that show outstanding characteristics, and some from other beekeepers with stock with an interesting strong suit but are from a different location, background, and gene pool than yours. Consider swapping good stock with your neighbors; it should be practiced much more that it is.

■ Single grafting, transferring a larva from frame to cell cup only, is most common, but priming a cell cup (use only plastic cups, which are sterile, or make your own wax cups with your clean wax) with royal jelly diluted 1:1 with sterile (distilled) water before placing the larvae in the cups is sometimes used. Double grafting is also an option. Double grafting is when a larva is placed in a cell and the cell is placed in a special starter box for about twenty-four hours. The larva and remaining jelly are removed, placed in a second cell, then moved to the regular starter box. The additional royal jelly is believed to give the new larva a nutritional boost. There have been concerns about contaminants in imported royal jelly, primarily antibiotics, but other chemicals as well. Exercise caution when purchasing this material for double grafting.

Cell bars with plastic cups await grafting

■ Grafted cell cups are moved to a young bee-rich, food-rich, queenless starter colony for a short (twenty-four to thirty-six hours) but intense period of time. Starters generally have lots of young bees, and lots of open larvae, usually in two or three boxes, with the queen cells placed above a queen excluder. (Queen cells are first moved to a very strong and well populated [with young bees] queenless colony because acceptance is much better than in a queenright colony. But queenright colonies raise better cells, once accepted. You can't do both in the same colony as successfully as when using two separate colonies.) The bottom two boxes in a queenless starter colony are loaded with honey, pollen, and brood. The bees in these colonies are feeding an exceptional number of larvae downstairs, plus anywhere from 15 to 100 queen larvae upstairs. You can't have too much food or too many young bees in these colonies.

■ After this intense enrichment period of about thirty-six hours, the barely started, but accepted, queen cells are moved to an even more populous queenright (with the queen below an excluder) finisher colony that is continually repopulated from dedicated support colonies, and is rich in both food and young bees. This step ensures the now-accepted queen larvae continue to receive the best possible care and exceptional nutrition. Fewer queen cells are moved into each finishing colony because they are going to be there longer and require more resources from this support colony.

■ Some beekeepers, many of whom are very successful with this technique, use a frame that holds three cell bars in the finisher colony, each bar progressively older than the one below it. The bar of cells on the bottom is the youngest and closest to the youngest bees in the box below, thus getting the most needed care. When the next bars are moved out of the starter colonies and added to the frames in the finisher colony, the cells in the top bar are capped and ready to harvest, the bars in the center are moved up a notch to the top of the frame, the bars in the bottom are moved up a notch to the center, and the new cells are added on the bottom. It's a continuous rotation process that chews up young bees by the millions, and raises fantastic queens by the thousands.

■ Nearly sealed or sealed queen cells go in one of two directions. Sealed cells are harvested and inserted in nucs, mating nucs, splits, or established colonies for mating. Or they may be held for a short time in a temperature-controlled incubator until just before they emerge, or until they do emerge and can be moved to a queenless mating colony and installed as virgin queens. Sometimes they are sold as cells right after they are sealed, and sometimes the virgins are allowed to emerge, usually directly into cages in incubators, and sold as virgins.

■ Mating yards for these virgins must be in areas saturated with the drones of your choice. You may be mating narrowly focused queens with a broad spectrum of drones, or mating a broad spectrum of queens with narrowly focused drones, or some combination of these, depending on your goals.

Whichever avenue you choose, drone stock is as important as queen stock. See "Raising Drones" (on page 98) on controlling the drones in your area.

Requirements of the Starter/ Finisher Method

Raising queens with starter/ finishers requires an equipment- and bee-rich operation: starter colonies, finisher colonies, support colonies, mating colonies or nucs, a grafting house, several drone colonies that are close but not too close to the mating yards, extra beeyards, and more. Plus, it requires lots of bees in all of these colonies. You can never have too many bees of the right age in too many strong and healthy colonies to raise queens this way. It is a bee-rich, equipment-rich, time-rich, labor-rich, space-rich, food-rich, luxurious way to raise queens. If you provide all of that, it's a good way to raise queens.

No Shortcuts

But here's the kicker. If you cut corners you begin to cut quality. Which stage of this operation is least important: the starters with lots of bees, the finishers with even more bees and even more equipment, the queens you grafted from, the grafting process itself, or the drone yards? You cannot ask that question. Each of these steps is equally critical. Shortchange one and the whole castle tumbles.

It is an enormous investment in time and equipment to produce queens this way. Producing enough queens for your operation will tie up many colonies, and these colonies could be making money in a pollination contract, making honey, or making splits to sell—all of which would be generating cash at a time of year when cash is usually short. You have to want to raise queens the right way to get the queens you want. Remember, you have been buying the "shortcuts" of other queen producers for years. This is why you are tired of the crap queens you have to buy every year and have decided to raise your own.

Producing a Surplus

Once you have this bee-rich, equipment-rich setup prepared, why not raise more queens than you need for your operation and share your extras with the beekeepers in your area? They can stock their colonies with your queens (and thus produce drones that will act as drone source colonies for you later, plus cast swarms so the feral colonies in your area throw the drones you want, too). You can also use extra queens for the nucs you are producing each spring or summer for sale, for replacement colonies for your operation, or for expanding your operation.

Your queens in your splits is a great way to handle spring swarming with your very populous colonies, or to generate cash at a cash-lean time of year. Because you are moving less for pollination and not at all for honey production, spring income needs to be looked at, and if pollination income declines, nuc and queen sale income needs to increase to accommodate the difference.

Expanding Your Queen Production Capacity

Producing queens is an important part of your operation. It is the only way you will ever get the bees you need to succeed. If you only produce the queens you need for your operation, when you are done your cost per queen will be high because of the relatively small number of queens produced, but acceptable because of the quality. However, if you produce more queens than you need and are able to sell some to recoup some of your costs, so much the better: Your cost in producing some small number of queens is minimal.

But what if it turns out you're really good at this? That you enjoy the stress of dealing with bad weather, fussing with erratic delivery systems, both consoling and educating customers who at times appear to be dumber than bricks, and struggling every spring to find good help that you just have to let go at the end of the summer? What if your nuc and queen business far surpasses your honey production goals or your candle making goals? If you create locally produced queens that do what you say they will do, arrive on time, are alive, and are backed by your word, the world will, I guarantee, beat a path to your door ... and I'll be first in line.

If you enjoy the challenges of the job and see a steady and secure financial future in raising and selling queens, beekeepers need you—desperately. Please, consider this a golden opportunity to excel in a field few others have made work.

Starting a Starter/Finisher

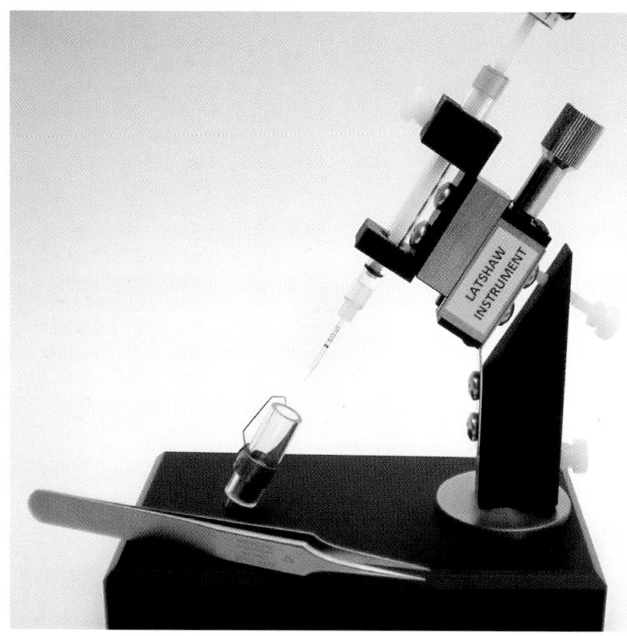

Inexpensive instrumental insemination devices are available, making this an option for producing the exact breeder queen you want. The extras—carbon dioxide (CO_2), cages, and the rest—are important, but this is the heart of it all. Newer models are far less fragile, and make it easier to loan, or rent, out.

This yard holds the colonies that have the breeder queens, and all the starter and finishers and several support colonies. This particular operation uses a single three-story colony for the starter, which feeds five two-story finishers. Raising queens using this technique requires a lot of equipment, but produces a lot of queens.

The beekeeper is standing in the back of the U-shaped starter/finisher "unit." The colony on the right in the back is part of the queenless starter, with the box on top being moved to sit on the rest of the starter. A queen excluder separates the top box with the queen cells and the bottom two boxes. The other five colonies are queenright finishers, with excluders between the two boxes of each.

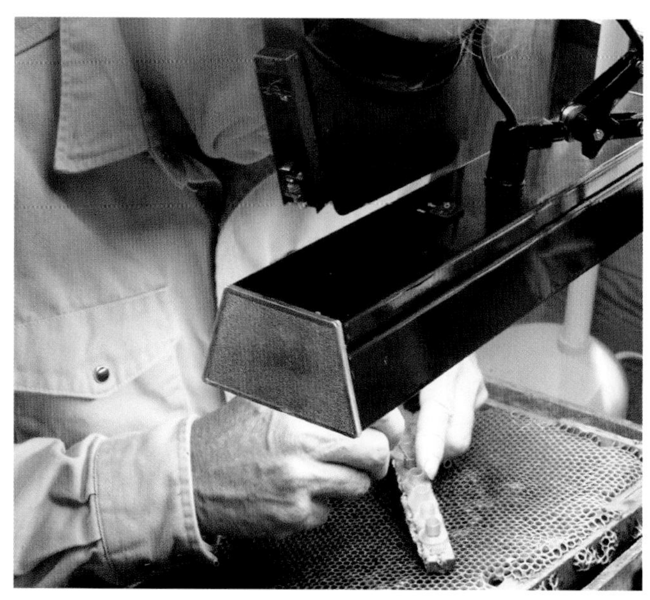

Larvae are removed from their cells and placed, dry, in plastic queen cell cups. These cups are easily removed from the bar they are sitting on now. Each bar holds 19 of these plastic cell cups. When all 19 cells contain larvae, (the process of placing larvae into the cups is called grafting) the bar is added to a standard frame, which has been modified to hold two or three of these bars.

After thirty-six hours the cell bars are inspected for "take," that is, whether the starter colony provisioned each cell adequately, and began to build the waxen cell on the edges of the plastic cell cup. If the starter colony has a problem, it will be noted at this stage and can be fixed quickly so the flow of cells through the system does not stop.

The Cloake Board Method

Increasingly, raising queens using a Cloake board, so named after the inventor, is becoming common with midsize operations because it requires less equipment and labor to produce queen cells or even mated queens. It is easier for a small operation to produce a lot of queens on a tight schedule. Plus, the equipment is easy to set up, uses minimal resources, and then converts back to regular equipment in a heartbeat to be used for honey production or pollination, or with all the bees you've added, it can be used to make three, maybe five splits.

A Cloake board colony has two parts—a queenless starter top half that's young, food-heavy, and bee-rich, and a yet richer-young-bee, food-rich, queenright finisher bottom half.

The two boxes are separated by a modified queen excluder that includes a metal sheet used as a full colony separator. This sheet can be slid in as needed to distinguish, and separate, the top starter half from the bottom finisher half.

This unit—excluder and removable sheet—is considered the "board" part of the Cloake board technique. When the sheet is removed, the whole colony can be a finisher colony if needed, with a queen below the queen excluder, and the developing queen cells above. When sealed, queen cells are ready to be removed to go into an incubator to let the virgin queens emerge or are placed in a mating nuc to emerge or sell. The biology for this is quite simple. Like all queen producing units, no matter the exact technique, the colony doing the work needs to be properly manned. Experienced queen producers begin this process months in advance, ensuring the colony is safe, clean, and run by a productive queen. Weeks before the rearing process begins, the beekeeper adds open brood to the population from support colonies to ensure there is an abundance of nurse bees to care for the queen cells to be added. This is the right number of bees of the right age in the right place at the right time at its best. This colony, if prepared correctly, is right on the edge of swarming. But, not quite.

A couple of days before grafting set up the Cloake board on the colony to be used, being sure to check both boxes you will use for a queen. The biggest problem producers have with this technique is that there is more than one queen—often one in each box. Make sure there is only one, and that she is in the bottom box when you apply the board.

Apply the Cloake board so that the entrance is in the rear of the colony. Make sure the entrance is left open.

The colony is ready. You can add a super of honey as insurance, and if you have one, it is advised. There should be a lot of bees in these boxes, and three or four days of spring rain will challenge the food reserves.

Both the top deep box and the bottom deep box have entrances. The top is supplied by the modified board; the bottom uses the regular entrance. Some producers reverse the top box just before beginning, so that the top entrance is facing the back of the colony, and open them both a day or two before the grafted cells are added. When the cells arrive, they close the bottom and open the top entrance to increase the bee population above the excluder. But these are forage-age bees and offer little assistance in queen rearing. Because of this many producers now leave both entrances facing the front, closing neither. It simplifies the process and with as many bees as these colonies have, it helps increase ventilation and reduce swarming. A couple of days before grafting, set up the colony with the Cloake board, making sure the queen is below (and making sure there is only one queen). In the top, put in a frame of open brood, a frame of pollen, and one of honey, and maybe a feeder. Have the entrance for the top box facing the back of the existing colony.

Below, keep the queen, and make sure there's honey, sealed brood and empty comb for the queen to lay in. Make sure all entrances are open.

The day before the graft, insert the slide, make certain the top entrance is open, and remove any uncapped brood from the top box. This causes the nurse bees to have plenty of food ready for the queen cells, to come the next day. Leave space for cell frames. Add nurse bees from support colonies. You can't have enough young bees. Beef up pollen and honey supplies if needed. You don't need to close the front entrance for the bottom box.

The day of the graft, remove any open brood and place cells in the empty spot left from the day before. This top box is now your queenless starter colony, completely separated from the bottom box by the

Insert the slide the day before the graft, and make sure the top entrance is open. The top entrance will be the only exit for the bees above the Cloake board, and the only means of ventilation.

The day of the graft, add a frame of pollen and a frame of honey, and remove any frames with any open brood. Leave room for the frame containing the grafted cells to be added later.

slide-in sheet and well stocked with engorged nurse bees.

A day after the graft, remove the slide and close the top entrance. Check again for a (missed) second queen. This reunited colony is now the finisher colony. Keep it this way until the cells are capped and then move the frames to holding colonies above a queen excluder.

Harvest on day ten or so to sell, insert in nucs, or requeen your own colonies.

This technique has a lot of advantages for an operation that isn't equipment rich, or is operating on a schedule that needs equipment for multitasking during the season.

The photos on pages 106–111 provide an overview of how both of these techniques work, but if you find one or the other to your liking I urge you to research the techniques of grafting, feeding, moving, and the rest in more detail so you have a good understanding of the techniques you think may be best for you. There is a wealth of resources available for queen production instruction, biology, and techniques at the end of this book. All are readily available and all are extremely useful. Check out as many as you can.

Place the frame with the grafted cells in the space you made available when removing the remaining open brood. There should be pollen on one side and honey on the other side of this frame. This top box now has its own entrance, in the back, is isolated from the box below by the slide-in sheet, and is queenless. It is your starter colony.

The day after the graft, remove the slide and close the top entrance. The colony is now completely queenright, with the queen in the bottom box below the excluder. This is now your finisher colony.

You may have already been involved in some level of queen production before now. Most beekeepers at this stage have dabbled in the art, but the dedication of equipment and time may have been more daunting and expensive than seemed worthwhile. The biological principles and requirements are straightforward. And by now I suspect your perspective for both equipment and time commitment has changed relative to the cost of poor-quality queens in both the short and the long run. The expense of putting two or three queens in every colony every year needs to be reviewed. Even inexpensive queen cells, when buying several for each colony, add up when considering the time spent on installation, resource management, and lost production. And by now, time management is probably your biggest issue.

The commitment of equipment means more than just taking two, four, or eight colonies out of the production line. These must be your best colonies, not your worst colonies. They must be your cleanest, completely uncontaminated colonies. The wax in these colonies should have never had contact with in-house pesticides. And any stored honey and pollen must be as pure as the driven snow.

After two days check the cells to see how the colony is doing, and to make certain you haven't missed a queen. They should be well on their way to being capped, be well-provisioned with royal jelly, and have lots of bees working on them.

No matter which technique you use, raising queens when bees want to raise queens is always recommended. Waiting a bit, until the weather has really settled, when drones are really abundant and swarming issues have already been resolved, is strongly recommended. These mid- to late-summer splits can be used to thwart varroa populations, can provide strong colonies for overwintering, and are an excellent source of income. Take a look at late-season splits later in the book for additional information.

A strong finisher colony, whether it is a regular finisher or a Cloake colony, will produce a lot of well-fed healthy cells.

Background Check for Drone Colonies

There are two ways to look at reducing inbreeding in your operation. One way is that you can provide fewer of your queens to area beekeepers, thus hoping for diversity to happen naturally. That is, if there are enough bees out there. You probably have a feel for that already. If there are lots of beekeepers buying packages every year from various sources you'll avoid inbreeding, but what are you receiving in return?

If you flood the area with your queens, what then? If you graft from a diverse line of breeders, spreading them around shouldn't be a problem. An insurance factor is to provide a few of your drone colonies that aren't a part of the queen's gene line. Something different, but something you like that you wanted in the mix anyway. This is one way to introduce new genes into your operation.

Until you grow your operation big enough, this isn't something to lose sleep over, but do not forget that it will enter the picture eventually. And always watch for that shoddy brood pattern—it will tell you what's going on.

The message the bees have been sending us for the past two decades is pretty clear. If their nest is contaminated—with chemicals you add, or chemicals the bees bring home from the killing fields of modern agriculture—the results are predictable. Colonies die or they perform far below our expectations. This is evident in the queens that are commercially produced; at best, they are consistently erratic in production and exceptionally short-lived. There is a better way, and producing your own queens is the best first step to return to beekeeping the way it should be.

Making Better Beeswax

You must plan ahead by several months, maybe even by a whole season, if you are going to provide enough clean comb to outfit several full-size breeder homes, starter/finishers, mating nucs, and drone colonies. You can't just purchase clean foundation at your nearby big-box retailer, and no matter what technique you use you'll need clean, drawn comb. Here's where compromise can be lethal. Do not think for a moment that you'll get your first batch of queens started and then work out where to get better wax. It's a death-trap decision for your bees and any queens you will try to rear.

What I believe to be the best way is not the first choice of many, but for my money and time it's the easiest way to go. Use unwaxed, plastic foundation. Add your own clean wax to the foundation with the sponge brush technique. You can add all the wax that's needed, so be generous. Well-waxed foundation will be drawn just as fast as sheets of beeswax foundation and the amount of cross comb and missed spots will be the same. To me, it's a no-brainer.

I can hear the pure and natural voices chiding me in the background, so if you can provide some, use starter strips of your pristine wax in wired but foundationless frames. This allows the bees to fill in the rest with new wax. This is slow, it can be messy if there's not a flow, and it can really be a pain in the bee suit. But it works, and the top-bar beekeepers have it down to nearly a science. (For more on top-bar beekeeping, search the Internet, but practice caution with the information you receive there.)

Another way to ensure top-quality beeswax is to make artificial swarms from bees you have now that will draw comb like mad, either on your starter strips or on homemade foundation (made by you or someone else with a roller). Either way, this will take some time to organize. You'll need lots of wax, that's for sure, and rollers are expensive. Consider borrowing one or contracting someone who has one to make you some from your own wax. Find someone who can help.

The truth is that there are no shortcuts to a cleaner way of life. Once you're geared up with any of these you can produce enough comb for a moderate-size queen production operation—say, 200 to 500 queens in a short season—in three colonies in three or four weeks. For more, the plastic foundation method looks better because you will need hundreds of frames.

But whichever technique you choose, keep producing and harvesting and using only this wax so you can eventually recomb your whole operation. Use swarms, splits, huge colonies in a honey flow that are just begging for a place to put more wax so they have a place to put more honey.

But be warned. Don't put these nearly sterile frames on a hive that has old, dirty wax already inside. Bees move wax all over a colony, and with it all the junk you're trying to avoid. Make it all new wax, or all old wax, in a hive. Borrow a page from the top-bar hive folks and get good, clean comb in frames that you can now use for the next three to five years.

The Working Class

(Or the Bees That Beekeepers Keep)

You've managed to raise perfect queens. Beautiful, aren't they? They have a dusky bronze tail fading to gold at the very tip, long legs, and crystal-clear, iridescent wings. Big as a house they are, fat in the middle, and so very long. Those slender golden front legs go all the way to the bottom of a cell. If bees had centerfolds, any one of these beauties would be Miss June for sure.

And, now that the inside of your hive is as clean as you've been telling everybody it was all along—there's no toxic residue, no pest, no disease, no nothing that's harmful to you, your crop, or your bees—there will be more bees than ever, living longer and living healthier.

Add to that the fact that all those well-bred, well-cared-for bees, living in that pristine environment, have all the food they want, whenever they want it. This is a beekeeper's dream: lots and lots of bees, right where you want them, when you want them there, and all as healthy as can be.

So what now? For starters, you want to avoid having the wrong number of worker bees. More specifically, you want to avoid jamming the right number of worker bees in the wrong, too-small place.

What exactly does that mean? We tend to take workers for granted. Queens, they're in a class of their own and deserve special and continued attention. Drones are special, too, produced in droves in their holding colonies during mating season, or produced by the pound as bait for varroa, but reduced to third-class citizens when times get rough any time of the year, and especially late in the season. But workers, they're sort of like factory folks: They show up on time, punch a clock, do the job they were hired to do, but can change on a dime if they have to, and the tasks they do evolve as seniority increases. They punch out at the end of the day, and at the end of their time they move on. Workers need to be predictable, reliable, efficient, industrious, and more or less interchangeable. These are the bees beekeepers keep. Take care of them, and they will take care of you.

The Big Three: There are the "big three" activities that workers do that will mess up your business if you let them. You know what they are: springtime swarming, everything to do with honey production, and dealing with pests and diseases.

Preventing Springtime Swarming

Some small percentage of your colonies will probably swarm every year no matter what you do. I look at it as a cost of doing business. It always comes down to that time and money thing. And losing a swarm from a strong colony, as expensive as that is, is way, way better than cleaning a deadout. It's better because you still have, at the very least, only a slightly wounded pride, but a pretty strong colony left behind that will offer some reward down the line (if you attend to it before it starts throwing swarm after swarm).

If you adopt the following practices, you'll minimize swarming without a lot of cost. (Most years, sometimes.)

You already know that you have to be way ahead of the curve to manage a large population of bees in a box at the time of year they are programmed to go. The irony is that everything you do as a beekeeper, you do to produce lots of bees, and then once you have a lot of bees, everything you do is try and keep all those bees from swarming. God, it appears, truly has a sense of humor.

Before Spring Arrives

We'll discuss many management practices you can do the previous season that significantly reduce spring swarming, but if you didn't, or don't do those, spring swarming is much more likely to occur. But even well-prepared colonies will flirt with throwing a swarm. So here's what you should be doing with most of your colonies, just to make sure they don't end up in the trees.

Your preventive measures start while it's still winter. The first inspection for food should be so early that you wear a winter coat, a winter hat under your veil, and boots if you live where there's snow or mud. You don't want to do a full-blown exam with frames all out, but you want to know that there's still honey in the frames closest to the cluster, and that the bees aren't at the very top with nowhere to go and nothing to eat. If you can't see honey close to the cluster and there's honey to move, then move it. If there's no honey to move, then feeding is in store—and you goofed last fall, but maybe you can fix it.

The two things that will mess up your honey crop are swarms and bad weather. You have a lot of control over the first, but you can only hedge against the second.

Making a Candy Board

1. For each board, pour in 15 pounds (6.8 kg) of sugar and a 1-pound (0.5 kg) jar of corn syrup. Add 4 cups (946 ml) water, heat, and mix.

2. Bring the mix to 240°F (116°C), turn off the heat, and then let cool to 180°F (82°C).

3. Fill the board about three-quarters full, let cool, and harden. Place on the colony in place of the inner cover.

To make a candy board, secure a heat source (here a propane heater), a tub large enough to hold everything, and the boards, which you have to make.

Early-Season Feeding

Feeding very early in the season isn't difficult. For carbohydrates you can use candy boards you made during the winter because you knew some colonies were going to be short. You can use bakery-grade fondant—that easy, but expensive, mix of HFCS and sugar. The candy board replaces the inner cover so it sits right over the top bars, and fondant is laid right on the top bars, so you may need a shim to make room above the top bars. Upper entrances, wrapping, ventilation, and some still-cold weather configurations will need to be considered. Getting food in is important.

Protein supplement can be mixed into the candy board, and even the fondant, if you want to melt it and mix it in and let it firm up again. About 5 percent of protein supplement by weight can be blended in: That comes to about 1 pound (0.5 kg) of protein mix for a typical candy board with 20 pounds (9.1 kg) of sugar. Use this formula to calculate your own quantities of fondant and supplement. (The fondant supplier I use sells 50-pound [22.7 kg] boxes that I cut into 10-pound [4.5 kg] slabs, so I add a half pound [227 g] of protein at a time—which is hardly worth the effort.)

Make sure there's enough food, more than enough of the right kinds of food, way more than enough so you can sleep those cold spring nights in peace.

Fondant can be prepared many ways. If emergency feeding a small amount, fondant that is prepared ahead of time and frozen in a plastic bag works well. Cut the bag open and the bees will be all over it. For more feed, simply slice off a chunk from the block you purchased, at least 5 pounds' (2.3 kgs') worth, and lay it on the top bars. Use a shim to make room, replace the inner cover, cover, and you're done.

If you need to feed your bees carbohydrates, they probably need protein too, way more than less than a pound of supplement. Five pounds (2.3 kg) is adequate, but you can provide them with up to 10 pounds (4.5 kg). Get patties, or mix powder with enough sugar and water and even vegetable oil to keep it from drying out. Make a king-size patty that they can sink their mandibles into. If your bees are hungry, you must feed them enough of the right kind of food. Consider supplying vitamins and minerals: Some supplements have them, some don't. Don't short-change your bees now because you got lazy or got behind last fall. Recall that it is the brood being fed. If there is a lack of pollen, adult workers sacrifice protein from their own bodies to feed the young—for a while. They'll quit when it seriously threatens their life and then they throw out the larvae. But the adult will have a shortened life span at a time of year the colony needs all the bees it can muster. Cheat them now and it's the early honey flow they—and you—will miss.

Opening and Repacking the Colony After Feeding

Early-season feeding means opening your colony, which means unpacking all the winter preparations you made (see chapter 5). If you've insulated your bees with a corrugated box, feeding is simple. Add a shim under the insulated inner cover, put the food right on the top bars, recover, and close the top (but put some support above the food so the top doesn't sink into it). But if you've packed the colony with lots of insulation and coverings and tied it all together, you have more work to do. What are the chances the bees need help? If you didn't leave all the food you were supposed to, bite the bullet and find out. Unpack, then repack, even if it feels like a hassle. Do not let the bees die because you were negligent.

What a cost. What a waste.
Do the right thing the first time. Make sure there's enough food, more than enough of the right kinds of food, way more than enough so you can sleep those cold spring nights in peace.

But the really big problem is that colonies that come through the winter that are even average for your operation will be in better health and have bigger populations than the colonies you are used to, and because of these factors will grow faster too. Swarming is definitely on your agenda now. This is especially true for colonies that were split a year and a half ago with a queen that's that old. But swarming can be an issue for those colonies that were splits last summer with brand-new queens. If not managed correctly large colonies are more likely to swarm than small colonies, and colonies with queens older than a year are more likely to swarm than colonies with younger queens. Mostly. You know the drill, or at least what the books tell you.

Leftover Honey

There's a downside to a colony having lots of food, and that's lots of food left over: visions of frames of honey, slowly, slowly crystallizing in the spring. Here's one solution: Stay up late one night and extract these frames as early on as you can, before they are solid. Visit every hive if you have to, but when you inspect and see that a hive has several frames that they'll never get to before the good spring flows start, pull them and extract. This process takes a lot of time, labor, and work, but that honey is worth three times what any sugar is that you just might have to feed, and even more valuable than what you've already extracted. And you won't get it at all if it's crystallized solid. Most of us have been there and done that. But you will need those frames sooner than you know, and they won't do you or the bees any good if they are still full of hard honey. Harvest them, bring them back to the honey house and warm them, and then extract, bottle, and place a special artisan label on it that tells the world what it really is. Winter Blend, perhaps, or maybe Snow Special—you get the idea. Sell it for more than your regular varietal or artisan crop honey.

And, what do you do with all the uncapped honey you left on last fall that's still uncapped and uneaten? If there's still good honey on the hive, then consider extracting—and feeding it back to colonies that need more food in the spring. Be sure to note if the uncapped honey has fermented, and if it has, extract and discard—don't bet the bees will eat it. And you'll have those frames when you need them. Drawn comb is priceless; fermented nectar has no value. If you can harvest, extract, and feed within a week or so it should work fine for you and the bees. But check for fermentation first.

Remember, the frames are the critical assets here. An empty drawn comb is worth all the honey you can get the bees to put in it as many times as possible. A frame with honey in it that the bees won't eat and you can't use isn't worth the wood and wax it's made of.

Reducing Congestion in the Brood Nest

Reducing congestion in the brood nest provides more space for brood, honey, pollen, bees. Be aware of the available queen pheromone per bee; watch crowding in the brood nest; have some honey, but not in the middle; the list goes on. Mostly, managing congestion is about having enough room in the right place for enough bees of the right age, with a queen that seems to understand all of this. And it helps to have a beekeeper trained to handle the rooming issues that come up every year at about this time.

The trick is to get most of the colonies at the same strength so you can treat (or plan to treat) all of them the same. Mostly, your colonies will be the same size, though you'll have some that are slower, smaller, and needing less of most everything except attention. Still, putting away the supers you planned to put on them at the end of the day is better than coming up a few short. Those smaller colonies can really take a lot of time, especially for so little income. Think hard about those dinks—they just aren't worth the time, or the money as honey producers, but they'll have a value later in the summer. Don't misplace your attention, or sympathy, on a colony that just isn't going to cut it.

Springtime Swarming

It's natural. It's what bees are supposed to do. But it screws up the honey crop. Don't let your bees swarm—easy to say, harder to do.

You need to figure out a few things about swarming when you have lots of bees in lots of colonies and not a lot of time to mollycoddle them. A casual list of choices to reduce swarming includes:

- The obvious but far too often overlooked solution—providing enough room for your bees before they realize they need it!
- Regular requeening using summer splits
- Equalizing colonies without requeening
- Reversing as needed, removing queen cells as needed, and no requeening
- Spring splits, requeening as needed, or self-requeening

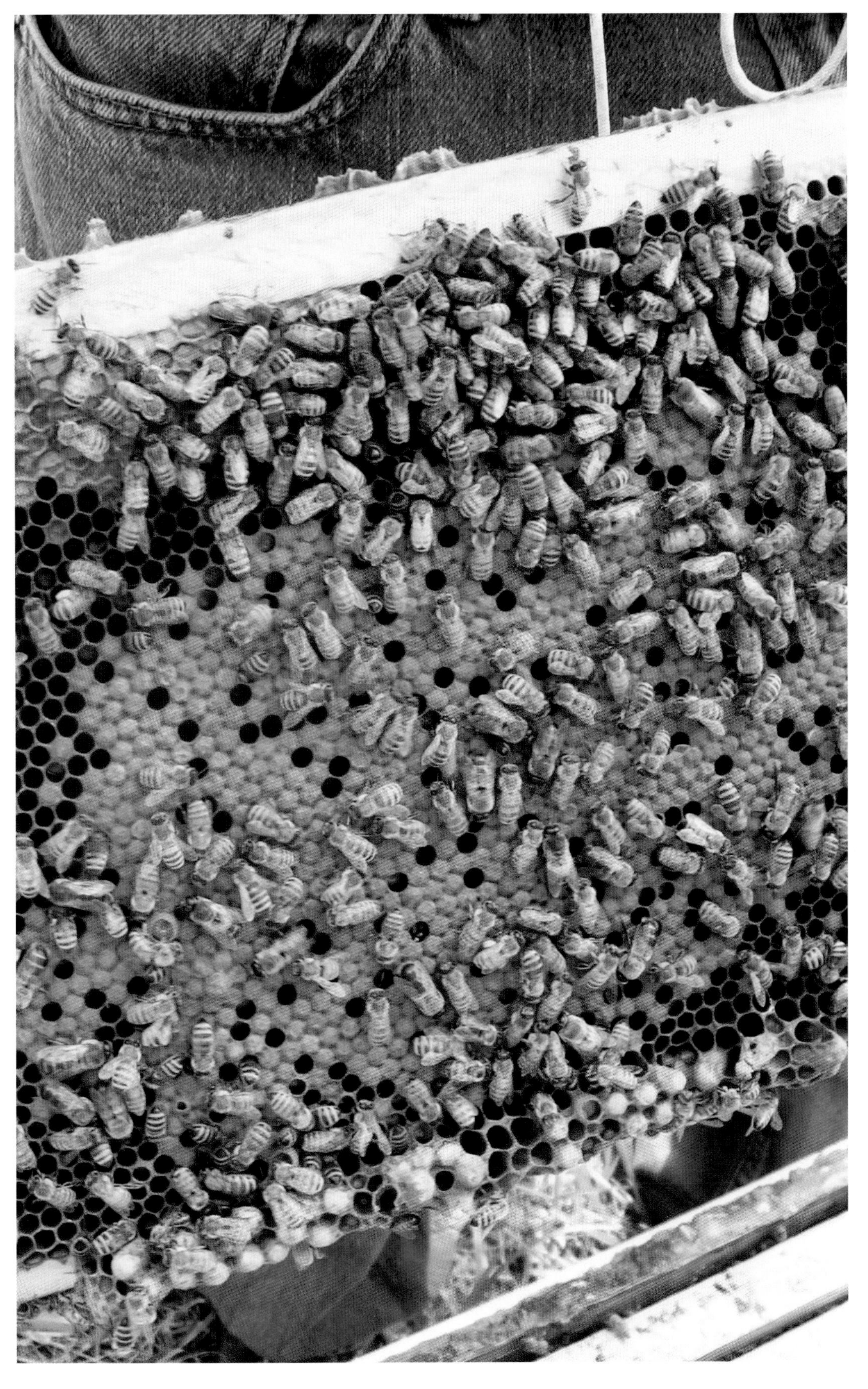

To equalize colonies that are not going to be split, sharing brood from brood-rich colonies with those that are a bit light is the goal. The colonies will be more alike than not the rest of the season, so you can treat everybody mostly the same, provided the queens are similar. Check to make sure weaker colonies catch up, and stay caught up. If not, check further to see why. These will be the prime candidates for summer splits.

Bottom bar down

Two sides of the same frame: How many adult bees and how much brood is on this one frame? Side one (bottom bar down) is about half full of brood, and about a third covered with bees; the other side (bottom bar up) is 90 percent covered with brood, and about half covered with adults. How many bees will this frame supply in less than two weeks, and how many frames does your colony have that are like this? It's no wonder they swarm when we aren't ahead of the population curve.

Provide Enough Room

In chapter 3 we looked at measuring brood to estimate the egg-laying rate of the queen. Here's another use for that skill that will come in handy early in the year.

We too often underestimate the amount of room bees need, especially in the spring. When you've been beekeeping long enough you get a feel for the population by the number of bees on top bars between boxes and how far from the edges they are on frames you can see, but that takes some experience. Even the old-timers will admit that for swarming and the honey crop, a quick look is better than a guess, but knowing for sure is better than a quick look.

But don't forget the time versus money quandary. Looking takes more time than guessing, and counting takes more time than looking—and swarms will take the honey crop.

How to calculate bees in a "quick look": Here are a few metrics to guide your calculations. There are between 4,000 and 4,500 cells on one side of a deep frame (about 50 cells tall x about 80 cells wide, depending on cell size and manufacturer), or 8,000 to 9,000 total on a deep frame. There are about 2,400 cells on one side of a medium frame (30 cells tall x 80 cells wide, depending on manufacturer), or 4,800 total. An adult bee covers two complete cells when she's on the surface of the comb, so when you

Bottom bar up

pick up a deep frame that's completely covered with bees on one side a single bee deep, you're looking at 2,000 to 2,200 bees. But if that side of that deep frame contains mostly sealed brood, you're looking at what will become, in less than twelve days, just over 6,000 bees, or enough adults to cover one and a half deep frames completely. And that's from just one side of one deep frame.

This quick look will tell you how many bees you have, and how many you will have in less than two weeks. You must now ask if there is enough room here for all the bees-to-be, plus as much or more brood as there is now. Also, where will all the nectar and honey go?

Recall that honey starts as nectar in cells. And for every final cell of honey that's made during a torrential honey flow, bees tie up three empty cells during the process of curing the nectar (some old-timers say four). But even if it's only two, you need more space than you think you do, just to accommodate a good nectar flow.

The plot thickens. You have an exploding population of adults, an exploding population of brood, an incredible influx of nectar turning into honey. You have copious amounts of honey and pollen already stored—and then you get four, five, ten days of rainy weather so that nobody flies and everybody's home. You didn't get supers on because you thought it's too early, I'll put one on, maybe two, and check again in a couple of weeks—which would be about a week after they've swarmed.

With enough room for brood, for adults, for nectar and pollen, and for honey ahead of time, but not too far ahead of time, you'll see fewer swarms. But you have to have that room before they even have a need for it. If not, the race is on, and all you do is try and catch them.

Equalizing Colonies, Discarding Dinks

Equalizing colonies is a good idea, but it's a lot of work. Your goal is to get the number of adult bees and each stage of brood about the same in most colonies so you can treat all the colonies the same the rest of the season. Having a lot of colonies in one place makes this easier, but more complicated at the same time, but tie your veil tight, put your gloves on, and do the job. It pays in the long run. You will save time by being able to treat most colonies the same for the season. Those that fail during the summer can be isolated for special treatment, joined to another to bring them both up to snuff, or dispersed throughout the apiary if past saving, helping others with brood or honey, or most often, used for summer splits.

Reversing

I always have trouble reversing, or switching positions of the two 10-frame deeps of a few of my colonies, or moving a couple of the three 8-frame mediums I mostly use for brood rearing and over-wintering. The bees are usually in the top two medium supers in the spring with lots of food left. I normally take the bottom medium, which is almost always empty, and

A major problem in early spring when there's not enough room is that incoming nectar will be stored somewhere—and that somewhere may be the brood nest if you haven't provided any other room. Every cell of nectar or honey in the brood area, or anywhere in the colony, is one less cell the queen has to place an egg. This is not good management. The brood nest should be sacred, and there should be lots of other room for honey.

put it on top, and add two more right away so they have a lot of room overhead. For 10-framers, I put the bottom deep, which is almost always empty on top, and add a honey super right away, which leads to lots of room right away. Sometimes they'll get all the way up to that honey super and put brood in it. I will put it on the bottom later. I seldom run into swarm cells the first time because I get going as early as I can—the weather notwithstanding. There have been times when we started on a pretty nice day, but finished checking in sleet with raingear on. It simply has to be done at the right time—before all that new brood is capped. No excuses.

Coupled with adding more room, reversing works to keep up with the bees' needs for space as long as they have food in the right place. Adding supers and reversing cuts down on how many supers you need, especially when using 8-frame equipment. You have to be ahead of the bees when it comes to providing space more often because they will fill all eight frames in a hurry. And you have to check again in less than twelve days so you know the rate of brood expansion.

An additional consideration to reversing is removing queen cells, every last one of them. Sure, most of them are on the bottoms of frames: some on the bottom bar, some in the comb near the bottom, some in a cavity on the side bar nestled in tight with the comb. Ever find a queen cell attached to the inside wall of the top or middle super? You are more likely to find the empty cell after she and her court have left.

Removing queen cells is not a reliable way to prevent swarming. It requires far too much labor for so little success. There are better ways to avoid swarms. And you are not working with the bees when you are doing this. Not at all.

You have to look everywhere for queen cells if you are going to rely on reversing to keep your colonies from swarming. They are usually on the bottoms of frames, and tipping a super up to look there will find most of them. But not these, on the face of the comb, those on the side wall of a nuc. You would have missed them all, and then you have missed your swarmed colony.

Making Spring Splits

If you made your divides and requeened last summer after the longest day you've mostly removed the urge to swarm from your colonies. It was long, hot, and heavy work then, but those grand colonies are a joy to work with in the spring. But some of those colonies are really big, and really need some attention or you might lose them. You would then miss out on having two producing colonies instead of one.

Some didn't get split last summer, naturally—and time, tide, and bees wait for no beekeeper. But spring splits work; tradition and old-school teachers still support that task. The biggest challenge is being able to get the queens, whether you are using mated queens, virgins, or even queen cells, to fill those queenless boxes. You probably don't have your own yet and what you get from elsewhere isn't what you want. And every day you delay means a day less of a honey crop and a day closer to a swarm.

Two decisions await those colonies: Raise their own queens or have one thrust upon them.

The Queen You Know

The trouble with queens you don't know is simply that: They are queens you don't know. Remember the bell curve we discussed in chapter 3? Some are great, most are average, some are dead, and a lot of them go missing after a couple of weeks. Supersedures are so common as to be expected. And though the colony's overall performance is

A twenty-queen shipment very early in the spring will supply your early spring splits, but recall that bell curve of queen quality: Four will be wonderful, four will be dead, and twelve will be average.

Sealed brood only can signal a problem. The queen has quit laying because she has been injured or killed and the colony didn't have young enough larvae to make supersedure cells, or the colony is in swarm mode and the bees have quit feeding the queen to slim her down so she can fly. In either case, this colony is headed for a major change, and you missed the boat. Check for cells and a queen, and get these bees a queen—by splitting and requeening the split, or simply requeening this colony (this is always the last choice, though, because acceptance is seldom a given).

From three or four strong overwintered colonies, you can create six or eight, or even ten, smaller colonies, then requeen them with cells, and let them build on the early honey flows. They will probably be strong enough to make a crop later in the summer, and be in good shape for overwintering—or splitting again for pollination next spring or even for a summer split.

predictable because of that curve, you don't have a clue about their genetic background (other than their color). Also, any previous equalizing efforts among all your colonies go out the door due to queen variability. This is the reason we often babysit every colony, instead of dealing with pallets, yards, or operations—the obnoxious variability brought in by all those variable queens.

But if last year's procrastination has won, take those big, overwintered colonies and do one of the following:

Divide them so you don't have to worry about them swarming at all, knowing full well it will take all summer to get them up to speed for next year.

You can also barely divide them, so what's left will be honey makers.

You'll need two big colonies to create a single honey maker. When you're done, you'll have three that are nearly equal in size that still might swarm (especially the one with the older queen, if you leave her). They will still need to be reversed, supered, and worked on. But you'll get a crop from these if you're growing your own honey crop on leased land right next door.

Still, the queens cause the problems here: Where will you get them? You should have resolved that last summer, finding the supplier who has the genes and the queens you want in your operation and arranged for cells early in your season.

Friendly Reminder: Buying Used Equipment

It is easy to get caught up in expansion plans when everything is going just right, honey is coming in faster than you have boxes for, and money is coming in, too (for a change). The weather is good and everybody is your friend, to boot.

Equipment standardization has been easy to maintain until now because you've had control: same supplier, same-size boxes, same type of frames and foundation. But then you buy somebody out, that nice old guy in the club who was moving and who just wanted to be rid of his stuff. Maybe he threw in a little extra and now you have three truckloads of equipment that hasn't seen the light of day for ten years.

You are, naturally, reluctant to discard anything that still works, or can be fixed—you should hear the mantra of time versus money loud and strong in your head by now—so the equipment that still works slowly gets incorporated into your everyday stream. Soon, you have all manner of "stuff." Some works, some needs fixing, some you're saving, and some is just junk. Pretty soon, everything will be mismatched and you will have absolutely no idea what you have.

One way to address this problem: Use the equipment storage space at your leased beeyards to isolate oddball equipment.

There are several odd-size equipment manufacturers out there. Your frames won't fit the bigger, or smaller, boxes. And there's the 8-frame, 10-frame mismatch, and the homemade stuff that doesn't fit anywhere. Be careful when you buy.

Another sad tale is when you buy equipment, sight unseen, from a beekeeper who always seems to have a lot of honey, a lot of bees, and a lot of time. You've seen a couple yards, been to his place a few times, and you assume his equipment is usable. Turns out it's very ugly stuff. The only way to address this problem is to get out the marshmallows: It's time for a very expensive campfire.

Making Summer Splits

Creating summer splits is almost the entire future of expanding your beekeeping operation. Start small and slow at first, iron out the details spelled out below, and don't look back.

Summer splits are the colonies that are queened with exactly the queens you want, whether they are cells, virgins, or mated queens. These colonies will go into winter with the best queens and the youngest and healthiest bees possible. They will be a colony and a half by next spring, with a young queen unlikely to swarm, but they will make you work like a horse to keep up with them all season.

Plus, even if you requeened with a laying queen, let alone a cell or virgin, there is a mid-season break in the brood cycle that makes a similar break in the varroa cycle that only helps contain those creatures. Using other varroa reduction methods—screens, drone traps, and the like—you can reduce mite populations in the mother colony, and help keep them low in the new colony. And you can do all this without chemicals (but don't assume you have mites under control—always, always check).

Summer Stock

Making summer splits should be on your to-do list. By now, you should have colonies to draw from, and the bees should have lots of external resources to draw on to support both a new colony and the split, plus whatever additional resources, food, and support you provide.

Because your queens are similar in age and productivity, and you have previously equalized your colonies, you have a good idea of how much brood, honey, and pollen each has. The overwintered colonies will be strong if they haven't swarmed already (and you have made sure they haven't, right?). You'll know about how much you can remove from each to make these splits. You'll also know how many of these colonies are too old or too weak (other than sick) to keep and nurse for a honey crop.

Given that, when you start you'll already know about how many splits you can make, so how much equipment to set aside. Equipment should be standard—as standard as can be—10-frame boxes, maybe split in the middle with a board to separate into two 5-frames, or better room for 4-frame nucs to start. The equipment itself is less important than the way it is used. Keep that in mind.

The colonies of choice to make summer splits from in this two-queen operation are those that didn't take the queen in the first place and have languished since. These are the resources from which to make new colonies, rather than simply requeening a slow colony. Make at least two from each, and put in the best queen—and you'll end up with at least two buster colonies next spring from these. Don't mollycoddle a loser—redo and make winners from them instead.

The best keepers use their weakest colonies—still healthy, mite free, and disease free to supply most of the bees for summer splits. The general attitude is why keep a dink if the dink is caused by a failing queen? These are boosted by brood and maybe some more bees from strong colonies, especially colonies in swarm mode. They are relatively large with an older queen, or they are just too big.

Making summer splits is a great way to reduce swarms in those strong colonies while still keeping them strong. And it uses bees that otherwise aren't optimally productive from those weaker colonies. Plus, it keeps equipment full of bees early in the season, and most important it gets the splits going for next season with brand-new super queens of your choosing. These are made toward the end of swarm season so there are still drones, and food is both plentiful and nutritious.

Feeding a split is encouraged even if there is a good honey flow on. (With your plantings, encouraged in chapter 2, there should be.) Plus, feeding anytime you introduce a queen is recommended: This is one stress on your queen and your colonies you can avoid. By fall you should have as many colonies as you need with queens that are less than a year old and ready to overwinter. You will know what you'll have for pollination, honey production, spring splits to sell, or whatever you need. You can always plan ahead.

Creating summer splits is almost the entire future of expanding your beekeeping operation.

Breaking the Brood Cycle

One advantage often cited to making summer splits is the break in the brood cycle that is created when there is no laying queen present. What that means, exactly, is not concrete, but it should be looked at closely, especially in light of routine treatments for American foulbrood or wax moth.

When you pull capped worker brood from one of the overwintered colonies in late spring, the likelihood of bringing varroa with it is minimal. Varroa tend to favor drone brood, and after a winter of suffering through worker only, or no brood at all, the spring flush of drones is refreshing to a female mite. Very early mite trapping on drone brood, though labor-intensive, is smart. In colonies with large populations of both varroa and bees, the mites are supposed to rush to those new drone cells that the bees are building at an intense pace in frames you've provided. Most years, early spring is queen-rearing season, swarming season, and drone-rearing season. And that is exactly what you want.

Mated varroa females head toward the available drone cells, leaving the worker cells you are going to remove alone. You then remove these relatively mite-free worker brood frames to begin nice, clean new nucs. Leave the varroa in the drone frames, which you will remove in a few days when capped. Then, you reduce the varroa population in both colonies.

And there's more. While your queen is not laying, there's no brood being produced. This opens a window of about two weeks or so (if using a mated queen) in which you are waiting for her to be released, and start laying, and for those slow start eggs to get capped. If you use a virgin queen, that window widens to three to four weeks, depending on weather and other variables. For a queen cell, add at least another week. If you let the colonies raise their own queens, add still another week. In other words, you can allow a colony to have no sealed brood for mites to hide in and reproduce in from two to six weeks. That is a long time for a pregnant, vulnerable varroa mite to be exposed to the elements. Falling off the bees she is riding on, being dusted off with your powdered sugar treatments, or being exposed to any number of other controls is possible during this time. This is your best window for varroa control.

But wait, there's more. With no brood for as few as two weeks, and maybe as many as six, there is a whole generation of house bees to help out with the foraging duties if they are old enough. These bees can also stay behind and help with nectar storage housework. Since there are no kids gobbling up the profits, more honey than ever gets stored.

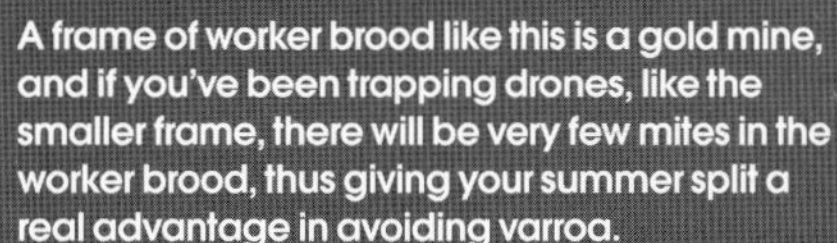

A frame of worker brood like this is a gold mine, and if you've been trapping drones, like the smaller frame, there will be very few mites in the worker brood, thus giving your summer split a real advantage in avoiding varroa.

All About Honey Production

"It takes lots of bees to make lots of honey": how old, trite, and true. The right number of bees of the right age, in the right place at the right time, and in the right condition will make you rich. Go ahead, laugh!

And you know, too, that a big colony (such as 60,000 adult bees, or ten to fifteen frames mostly covered in bees) will make a lot more honey than two (or even three) colonies with 30,000 bees each (that's only five or six frames). Remember, in either a large or a small colony that has good queens laying at the same rate, it still takes the same number of bees to take care of the brood in each colony. In the large colony, a greater percentage (and number) of adults are free to forage. With more foragers, there's more opportunity to find and exploit rich nectar sources. Plus, this larger number of foragers increases the variability of the resources found for a better, more nutritious diet.

Wait a minute. The question might be, "Well, okay, but you'll need more bees in the hive to receive all that nectar, right?" True, but not a significant number of forage-age bees.

This yard has had all the colonies equalized and the populations are set to take off. They have just enough room now to accommodate the adult population, but the frames of sealed brood are ready to explode, and more honey supers will be needed almost immediately. Being able to treat all your colonies the same saves an incredible amount of time when planning your work and managing an entire yard, instead of a colony at a time.

No matter what you do, how careful you manage your colonies, how far ahead of the curve you are, some colonies will swarm every spring, some in the summer, and maybe a few in the fall. It is what it is.

Look closely and you can see this sourwood tree is ready to burst into bloom; the racemes are ready and the flowers will open in a day or two. If the weather cooperates, these few trees will provide a wealth of exquisite honey. But only if your hives are ready—that is, lots of forage-age bees, and lots of hive bees able to receive all that nectar when it comes in. And the date these trees bloom every year is exactly predictable, whether in the mountains, plains, or wherever, if you use growing degree days to pinpoint the bloom date. There are no excuses for missing blooms anymore.

"Won't more bees eat more honey?" Some, yes, but not enough to put a dent in the storage budget. A colony with lots of bees has lots of bees for every job, so the same job (such as comb building or nectar storage) takes a big colony less time than a smaller colony.

Having enormous colonies is one thing, but having them at the right time is another. You must be aware of when the honey flows are, and where they are. Also, if you are renting colonies for pollination, you need to be aware of the bloom dates and length of bloom of the crops your bees will be on.

Reams of information exist on when nearly every plant in the world blooms, especially crop plants. Finding this data, no matter where you are, is relatively easy and you can plan accordingly. Be warned, though: At the time of this writing, the climate seems to have started changing, at least for the most recent seasons, so your planning might give inaccurate results if you are using historical data. Referring to growing degree day data is more reliable. Don't guess, don't look at last year's calendar—the key is to know what blooms, when, and why.

Refresher: Growing Large Colonies

Here are a few key points to remind you how to grow large colonies:

Last fall you made sure there was more than enough food for overwintered colonies that were headed by young queens.

For colonies that ate more than you thought, you fed those that were light very early in the season, and began feeding protein to all of your colonies to ensure they could begin raising brood as early as possible without having to draw on their own bodies for protein.

You stopped swarming in its tracks by: 1) overwintering strong colonies with young queens, 2) equalizing those colonies as early as possible to take the edge off the biggest and give the slightly smaller colonies a boost, 3) dividing overwintered colonies with old queens such that they weren't too small, and making honey-maker splits from them, 4) making sure everybody had enough room in the right place by reversing some colonies and adding room to others, and finally, 5) making sure they were so healthy that there was no slowdown due to pest stress.

For some you have provided a queenless break so they can focus on foraging instead of babysitting, while installing a new, young, healthy queen to lead the way for the next season.

You've done away with the dinks, either sharing their usable resources with other colonies, using their equipment for colonies that needed room, or using their resources for summer splits.

You have studiously avoided techniques that require a supreme amount of configuration changes and far too much work in favor of working with the bees and accommodating the environmental and biological requirements to keep them home and happy.

All this, and some still swarmed. Take heart: It's the cost of doing business.

Calculating Growing Degree Days

There's a simple technique for starting your own growing degree days (GDD) database, but first a ground rule.

Plants essentially don't grow if the ambient high temperature for the day is more than 86°F (30°C), or the low temperature for the day is below 50°F (10°C), so these two extremes must be noted in all calculations.

The basics of calculating GDD is:

Using various resources or media (such as weather websites or networks), find the high temperature for the day, add it to the low temperature for the day, and calculate the average by dividing by two. Subtract that number from the base temperature you are using, (which is 50°F [10°C]).

$$GDD = \frac{\text{Max Temp.} + \text{Min Temp.}}{2} - \text{Base Temp}$$

If the high temperature for the day was 84°F (29°C) and the low was 60°F (16°C), the calculation would be:

$$GDD = \frac{84 + 60 - 50}{2}$$

72 – 50 = 22 GGD.

Remember the extremes. If the high for the day was 90°F (32°C), the low 40°F (4°C), the calculations go like this:

86 (replacing the 90) + 50 (replacing the 40)

$$\frac{86 \text{ (replacing the 90)} + 50 \text{ (replacing the 40)}}{2} - \text{(minus) } 50 = 65 - 50 = 15 \text{ GDD}$$

The first question should always be, "When do I begin calculating this figure each season?" The simple answer is: a few days before the last frost date in your area. This annual event is available from gardeners, extension offices, most university Web pages, and governmental sources that feature agricultural information.

Alternatively, you can use any local program (check local university or grower organization Web pages for this information), and then refer to your records. The correlations are easy: When did maples bloom? When the local growing degree days reached a certain point. And they'll reach bloom when the growing degree days reach that point every year, rain or shine, sun, snow, or sleet, no matter what the calendar says.

Even though information for, say, apple growers, is being tapped, the data can be used to predict all manner of honey plant bloom dates. If you have a good history of those dates and access to past dates for agricultural growing degree days, the rest is easy. If not, a year or two of careful record keeping will give you the basics.

If you've chased land-based honey production for the land your bees are foraging on, you know the crops and the blooming dates and duration. If you are still at the mercy of the real world, then you need to be aware of the agricultural crops that bloom, the annual, perennial, and forest crops that your bees will search for. If you are in an urban setting, note the ornamental and street tree crops that may come to bloom throughout the summer. Once you have a handle on all of this and know the dates, then you can plan all the dates and times and places your bees need to be.

Measuring Growth

Remember the drill on calculating the egg-laying rate of the queen you have? (See chapter 3.) This is when it comes in very, very handy—that, and knowing your basic biology: How many days from egg to forager?

- About three days an egg
- About six days a larva
- About twelve days a pupa
- About ten days a nurse, house, and finally guard bee.

It's that basic, and it's that important. It's about thirty days from egg to forager. And she lasts as a forager for about two weeks—give or take a few days. If you know that your main honey crop blooms from, say, June 15 to July 1, then you must have the leading edge of the peak population growth curve in your colony on right about June 5, so you'll have as many bees as possible graduating from house bee to forager bee every day.

That means the house bee population needs to be large enough at least twenty-eight days before that crop blooms to care for the number of maturing and forager-ready bees coming on line every day during the bloom you are interested in.

To produce that quantity of house bees, you need to plan another thirty days ahead so there are enough bees to take care of those nurse bees. You always need to know when you need enough bees to take care of the bees that will be the foragers.

In short: For maximum performance from each colony, you need to plan to have the right number of bees in the right place at the right time. And, that "right time" is earlier than you think: long before the bloom begins. Think of it as a generational thing, as when raising queens. You need enough grandparents to take care of the parents of the foragers that will come online during the major honey flow.

It's tough to see this as an abstract concept, and it's tough to control given weather and seasonal restrictions. Commercial pollinators struggle with this all the time. They must time that leading edge of the population curve so they have and maintain a minimum population—usually at least eight frames of bees and brood—so all the bees they need are ready immediately, variables aside (still, too many commercial pollinators rely on their colonies building on the crop, hoping they won't get inspected until bloom is nearly over). Noncommercial beekeepers—the rest of us—are, on average, weeks behind that curve and are a generation late with our peak population. We haven't mastered the timing. We raise lots of bees right at the beginning of the crop with the intent of having the most bees possible for the crop, but too many of them stay home taking care of the kids.

So, what happens to all those bees when the flow is over? They are referred to as welfare bees: lots of bees eating all the honey they just collected. Come harvest, there is nothing left for the beekeeper. Planning beforehand will help that problem.

Preventing a Welfare Bee Population

- To not run out of forage: Plant additional crops for your bees to visit.
- Move bees to additional forage: It is an option, though not the best option. It's what beekeepers have done for years.
- Make summer splits: You have honey to spare, plus you can feed some to help the splits.
- Create a broodless period timed so that you run out of bees right about the same time you run out of forage. The timing is important, but not critical. You want it so that the last flush of brood matures at the beginning of bloom, rather than during bloom. This will help with varroa control also.
- You can also sell the summer splits, using the bees another way.

This is what you want to see three or four weeks before your honey crop comes due. Lots of bees, but more important, lots of brood. There will be enough bees to take care of the bees that will be foragers in these colonies. (Beware of bees on the camera lens!)

Pests and Other Problems

The author of one of the very best current how-to books available on a somewhat specialized beekeeping topic, *Increase Essentials* (see the Resources section), asked me to write a preface for his book when it was first published several years ago. His book covers basic instructions not covered here because it is intended for a more basic audience. The theme of the book was increasing the size of your operation. (Since you, reader, are already there, perhaps you should read this book to see if the author got it right.) Beekeeping operations worldwide have struggled for the past two decades to maintain colony numbers and health, and the simple fact is, numbers and health all come down to mites. And currently, it is just varroa mites. (I am certainly aware of the dangers of and damages caused by agricultural pesticides in the beekeeping world, and will not minimize them. Beware of corn and cotton and canola and sunflowers and almost anything that is grown commercially and needs protection from pests.)

You can link varroa mites to nearly everything that's wrong with beekeeping these days. And because they stubbornly refuse to succumb to most preventive measures or treatments, we will have to learn to live with them. Control, resistance, avoidance, protection: all the good IPM practices have some effect on their continued presence in our colonies. There are ways to implement these practices that won't put you out of business or contaminate your wax and equipment, and will protect your bees.

As I have repeated throughout this book, decisions about varroa treatment are an issue of time

It is no secret that producing honey bees that are tolerant or resistant to varroa mites is the only way forward.

This U.S. Department of Agriculture photo of a female varroa mite feeding on a pupal stage bee sums up the problem nicely. A damaged bee is susceptible to viruses and invasive diseases like nosema, has a dampened immune system, does not live as long as do healthy bees, and does not do as much work while she is alive. Control varroa mites, or go out of business.

These pure Russian bees show a high degree of resistance, or tolerance, to varroa mites. There are several ways they combat these mites: They have a fairly long broodless period, they are aggressive groomers, and they exhibit good nest hygiene, removing mites from cells before they emerge.

versus money, plain and simple. That being said, however, I don't mean to trivialize the necessity of absolutely being on top of the varroa problem. But treatment options and monitoring interpretations continue to evolve. Recommendations made here, now, will in all likelihood be outdated by the time you read this paragraph. My advice is to religiously follow current recommendations and guidelines from a reliable source that understands the particulars of your local conditions and management objectives. Then, use your best judgment and experience when deciding any course of action.

Rallying Cry: Varroa Mite Resistance or Death

It is no secret that producing honey bees that are tolerant or resistant to varroa mites is the only way forward. Everything else—drone trapping, brood cycle breaks, sugar dusting, screened bottom boards, and all the rest of the IPM tricks—is just keeping us afloat until true resistance is established in the worldwide honey bee population. The use of any of the chemicals in vogue is a fundamental crime against our product, our livestock, and our way of life.

There are at least two good reasons to get varroa under control.

Bee Damage

Varroa do direct damage to bees at all stages of development. They are the primary reason we say, "Take care of the bees that take care of the bees that go into winter."

Viral Infections

The varroa/virus relationship is silent and hard to see. In combination, the effects of varroa such as immune system breakdown and fundamental physical damage will make a colony susceptible to a host of lethal viruses. Also, viruses will capitalize on the effects of toxins on a colony other than varroa: low doses of pesticides, nutritional deficiencies, or nosema.

Signs of Hope

There are strains of bees that show strong indications that resistance to varroa is possible and profitable. Almost all of the lines that show this have some level of Russian strain mixed in. Over time, colonies of Russian bees consistently support fewer mites than other lines, all other things being equal, and show a tolerance for higher mite populations without succumbing to them.

Russian queen breeders (breeders, not queen producers) continue to improve their lines, and these bees are becoming more and more like the Italians and Carniolans we are used to. But they are not there yet—Russian bees are different in many ways:

- They are very resource oriented. They take advantage of food surpluses, and slow down during food shortages.
- They are extremely queen oriented, producing lots of queen cells during most of the season, ready to swarm or replace a failing queen if needed.
- Temperament in hybrids is erratic, but much better in pure lines.
- They are slow to begin brood production in the spring and wait until winter is long gone so food is available, missing early spring flows.
- Once they begin producing brood, they raise a lot in a hurry to take advantage of late spring and summer flows.
- They reduce, then quit, brood production when environmental surpluses dwindle during dearths in the summer and in the fall.
- They overwinter with very small clusters, consuming less than most bees, and certainly less than Italians.

To summarize, Russian bees have the following qualities: They start slow in the spring, produce fast once they start, will slow if there is a dearth, speed up once it's ovcr, may swarm more than you'd like, and are not predictably mild-mannered, but not aggressive, either, and shut down early in the fall, overwintering with small clusters requiring less food than you might expect from other types of bees.

There are other strains of bees that have had intensive hygienic behaviors selected for (sometimes too hygienic), but these have shown great promise in keeping the varroa populations in check. And there are lines that support a varroa-sensitive hygiene behavior (that is, removing brood that is infested with mites before the mites emerge) that is a bit more specific, which also shows promise in keeping these mites under control.

Every beekeeping magazine mentions these famous survivor bees: untreated for years with great success. And they do exist, though I suspect they rely on the influence of Russian or hygienic genes. It doesn't matter, however, if keeping bees alive is the goal. They work.

If you don't want to hedge your bets on resistance, there are a few good and safe options for active pest management.

This is an advertised survivor queen. You will find mites in her colony if you look hard for them, and there will always be a few when you do a sticky board count, but they don't overrun the colony, unless the colony is overrun with bees from an absconding hive that left because of a huge mite population. Survivor bees are good if the mite population reproduces normally.

Trapping, Avoidance, and IPM

Installing screened bottom boards is great for monitoring mites, but it does not contribute to reduced populations. The boards certainly help ventilation in the summer. They should be standard in every beekeeper's operation. (Incidentally, they were first suggested by A. I. Root around the turn of the last century for the express purpose of increasing ventilation in the summer.)

Trapping mites in drone brood is effective, but erratic, because drone production can be erratic. When spring is early, the weather is warm, and the colony is healthy, drones are generally plentiful. If traps are in early and drone production is consistent, many of the mites in a colony can be removed. But if drone production is less than ideal, due to poor weather, not enough food coming in, or a failing queen, this plan is not reliable. But if the colony isn't growing fast, neither is the mite population, so it sort of balances out. When it works it is immensely satisfying to rid a colony of mites simply by pulling a frame—but it is hard on the drones.

Your drone-producing colonies can be particularly hard hit with high mite loads. Switching frames of sealed drones—picked from colonies with the same genes you want but that have been in colonies that were treated, were used earlier for traps, or are resistant—to your drone yards helps, but it is time-consuming.

Breaking the brood cycle is effective, both as a mite control technique and in establishing new queens in your summer splits. It accomplishes four things, actually:

- Reduces mite populations.
- Helps raise new queens.
- Provides new colonies to grow or sell.
- Gives you something to do in the heat of summer.

Chemicals

Grouped together in this category are the organic acids, essential oils, and hard chemicals. It's the latter that have caused so much trouble by contaminating wax, but some are still effective. Control results from hard chemical treatments are variable because so much of the mite population has developed resistance to them. Organic acids and essential oil treatments are temperature dependant, so control is weather related, but when temperatures are not too warm or cool, they still work.

Organic Acids

Oxalic acid, though not legal to use in the United States at the time of this writing, is used in many parts of the world for good reason: It is safe for both beekeeper and bees, and it is effective in killing adult mites if environmental factors and timing are cooperative. It leaves no residue in the wax, the queen and what little brood there is escape relatively unharmed, you can time applications to fit management schedules better than with other treatments, and it is incredibly inexpensive.

If it is legal to use where you are, I recommend it as a treatment if:

- Using resistant bees isn't an option for whatever reason.
- You've inherited or gained possession of colonies that are infested and moving them to your operation would compromise your otherwise healthy colonies.
- You've experienced a varroa bomb (See next page).
- Varroa populations have gradually eased upward in a particular location due to pressure from a nearby beekeeper, a particular line of bees proved less resistant than you thought, or you are protecting some particularly vulnerable colonies.

The only issue with oxalic acid is that it is not effective on mites in brood cells. Thus, the less brood there is in a colony when treatment is applied, the more effective it is in reducing the total mite load. The problem is that waiting too long in the fall to treat may increase the amount of damage to the winter bees—leading to longevity problems for the bees over winter and into the spring. Avoid the very late treatment time if possible and focus on earlier season controls during dearths and before brood population rises dramatically in the spring.

Formic acid does not contaminate wax, and the new formulation is safer and easier on the bees and beekeeper. Once the new strip becomes widely available I anticipate it will be popular because, according to the current label, you can treat when there are honey supers on without harm to the honey. Early reports indicate there are still some queen and brood issues, but not serious enough to avoid the product. There are some temperature issues that can increase queen and brood damage and reduce control of mites, and the window of opportunity can be small some seasons. Backup plans should be considered.

A new organic acid varroa-control compound, (HopGuard), is made from hops, the flavor enhancer of beer. This food-grade product is applied as a liquid soaked in cardboard, which is eventually removed by the bees. It is safe and easy on your bees, hard on mites, and doesn't contaminate wax. If this kind of varroa control is appealing, or necessary, check it out.

Essential Oils

Essential oils are the next choice, but there is not much to say about these products. Moderate to excellent control with these is common, depending on external temperatures during treatment. They all work about the same way with essentially the same ingredients in different ratios. The primary differences are in the ease of application, the number of applications needed, and cost. There is some absorption in beeswax, so that has to be considered a downside, though the concentrations seem to cause few issues.

The Hard Chemicals

You may call me narrow-minded, but my advice is don't use them, ever, for any reason at all.

In conclusion, you have to control varroa, or you must have bees that don't mind them. Control consists of effective but somewhat difficult to use organic acids, a few inconvenient IPM tricks, and a few essential oil compounds.

The long and short of dealing with varroa is, in the short term, using the organic acids or essential oils. Long term provides a better picture, however, and that's general use of resistant bees. They are the only future we can conceive of. If you haven't, begin selecting your bees for tolerance or resistance on a local level. Bring in good genes, and keep bringing them into your program. Give your getting-to-resistant queens to everybody that has even a chance of having drones that might mate with your queens, and keep doing it every year until all the genes anywhere nearby are your resistant genes. It is the only way.

There's a thought that's beginning to grow that the main mechanism toward producing bees resistant to varroa is in the hygienic behavior that's so easy to select for. One scientist made an observation on the selection process for this that, if it turns out to be correct, will make life much easier for all of us.

It is thought that if the usual techniques of selecting for this behavior—that is, if killing brood with liquid nitrogen—is not possible for you, then simply keep selecting queens for breeders from colonies that do not have chalkbrood. It's a bit slower, perhaps, but much easier. Try it and see.

The Rest of the Lot

Varroa is the major problem, and we will deal with *nosema cerenae*, the runner-up problem, at the end of this section. The rest of the lot includes wax moths; small hive beetles; and bacterial, fungal, and viral diseases. They are all standard fare, and you have been dealing with them forever. If you need a refresher course on these diseases and how to cope with them, you can find this critical information in my beginner's book, *The Backyard Beekeeper*. It may be time to bone up on the basics.

Controlling Nosema

The nosema we contend with now is something of a mystery. When bees are infected with only this pest it seems to be less than lethal, and it can be handled, more or less. But as mentioned, when it is mixed with viruses bees many contract, the combination becomes quickly synergistic and lethal. The trick to offsetting the effects of viruses, it seems, is to control nosema. The research gives straightforward advice: Avoid it if possible. Remove spores from the inside of hives if possible. Treat with the only antibiotic available if you think it is necessary. Stop nosema, it seems, and you take the steam out of a lot of other problems.

One nosema treatment is not talked about much—the addition of a feeding stimulant to a colony's diet. Some beekeepers swear that when the bees are fed a megadose in syrup, it helps control nosema. Other beekeepers pour a strong dose directly on the bees, making them clean it up immediately. Some mix it in with protein supplement. Some do all three. The research is, at the time of this writing, minimal except for anecdotal information. Beekeepers report success with various treatments that are a hodgepodge of nutritional supplements, feeding stimulants, legal and illegal mite control chemicals, and even magical spells, but it's impossible to weed out the effective from the silly.

My experience is that using a feeding stimulant aids in microbial control. There is also good data to support that it helps, it seems, with knocking down viruses. It also helps with knocking down nosema, but it is unclear if it keeps spores from developing or kills them once they do. And even if none of these is as effective as some seem to think, in all cases treated bees eat more than untreated bees. Nosema seems to knock them off their feed and eating more and better food seems to help. A well-fed bee has a better chance of surviving than one that is food-stressed.

My advice: Use it as the label directs until someone changes the label. It can't hurt, and can only help.

Maybe, after all this, you won't relegate workers to the factory floor with such disdain anymore. These are the bees beekeepers keep. Take care of them, and they will take care of you.

A Varroa Bomb

When a colony of bees that is susceptible to mites is left untreated, the mite population will build up during the summer. When the bee population begins to decline in the fall and drone production ceases, mated mite females move to worker cells to reproduce. When a high level of worker cells become infested, the morale of the colony diminishes, and in extreme cases, the entire adult population, many with one to three mites adhering, will abscond, leaving a colony with only brood (and little of it), and no adults.

These adults will fly to a nearby colony that does not have a big mite population and move in, along with their mites. So a colony that had some level of resistance, or had been treated and had a low level of mites, suddenly has a huge population of mites. These mites immediately move into any worker brood available to reproduce, so in a very short time all the brood in that colony is infested, and many of the adults still have mites clinging to them. The colony, if not doomed, is certainly compromised. This is a varroa bomb.

Chapter 5:

Wintering Your Bees,

(or Baby, It's Cold Outside)

This much snow for months at a time ... that's winter in the extreme.

Unless you have your bees in the tropics or subtropics, I suspect you do not take adequate care of them during the season you call winter. Let me explain.

My definition of winter is when it gets really cold outside. There's snow, ice, and blizzards. Bees can't fly for weeks at a time, and for a lot of that time it's difficult for them to move around inside the hive too.

The days are short, more days are overcast than not, and on the rare days it warms up enough for the bees to take a cleansing flight or wander out to explore, there isn't anything at all to eat. And, for those that do venture out, many don't make it home (though it's mostly the older and perhaps the weaker of the bunch). That's winter, aside from what the calendar reads.

There are varying degrees of winter: Some winters are on the warmer or sunnier side, there are winters without snow, and some are even warm enough, often enough, (approximately every two weeks) that bees can be active for a couple of hours. In some regions, food may be available for some of the season. That's also winter.

If you live where your bees can fly all year, where there's lunch for the having anytime your bees want, and it's warm enough on January 1 in the northern hemisphere, or July 1 in the southern hemisphere, that you can wear shorts, you don't need what's here. But you should read it anyway, just to see how easy you have it.

But this may be as bad as it gets: no snow to speak of, though cool and overcast more than in summer. It's warm enough that bees can easily take routine cleansing flights. When they fly, however, there's essentially nothing to forage on. Outside access is available year-round and freezing temperatures are rare.

Winter Rules

Common honey bees with some yellow pigmentation originated in Africa (or so we're told): some from the northern part of the continent, and some from farther south. Though they continued to migrate north, some didn't move far, settling and adapting to the general area of central to southern Italy. That should tell you what kind of winter they are hardwired to handle.

Others continued north and spent enough time in the colder regions of Europe and Asia to become hardwired (and darker colored) to tolerate the first definition of winter. Your bees—yellow, dark, or mixed—are descended from these early generations, and genetically they have not changed much (or as much as we'd like to think).

Dark bees, like these Carniolans, have adapted better to colder regions and longer winters than their subtropical cousins. The Italians or other races still dominant farther south. Plus, northern dark bees have become more resource oriented because they do not have a ready food supply all year long, so must build rapidly when food is available, and reduce their population just as rapidly when food becomes scarce.

My Prime Directive, to borrow a *Star Trek* phrase, is that managed honey bee colonies should never die during winter. They should not starve. They should not die when it is too cold for too long. They should not die due to pests or diseases either.

Beekeepers are responsible for managing their bees and their colonies with respect and due diligence. Letting your bees sort it out for themselves with a "may the toughest bee survive" attitude borders on criminal neglect. If farmers chose that route with their cattle, homeowners with pets, or even zoos with their hundreds of animals, the powers that be would intervene and take control of these onerous situations. Yet beekeepers, since there is no threat of liability, are allowed (legally) to mistreat, undertreat, or not treat their bees so they do not have adequate protection from the elements, are not safe from pests and diseases, do not have enough food to eat, or some combination of these curses. Why would you let a colony expire for reasons of neglect?

Unnatural Habitats

Living in a box in the middle of a field is not a natural home for any of the bees we raise. So to not only survive winter but emerge from winter thriving, healthy, and ready to meet the next season, our subtropical bees need help.

When your combs are so dark that you cannot see sunlight through them, or they have been in your hive for three seasons, they are too old and too dangerous to keep. If you use plastic comb, scrape or, even better, use a flail uncapper to remove the wax and give the frame back to the bees.

The homes we choose for our bees are not the homes they would naturally choose. Feral colonies choose homes 15 to 20 feet (4.6 to 6.1 m) off the ground, with entrances at the top, usually in protected venues, and in tree hollows that are insulated against the weather and predators. A poorly situated, poorly insulated honey bee colony in this position needs all the help we can provide so it successfully comes out of winter.

Removing toxic wax is the single best thing you can do to help your bees, no matter the season. Rid your hives of the poisons, the toxins, the disease, and inoculum that is present in the waxen combs of your frames. If you are treating your bees for mites and diseases, no matter the chemical of choice, some amount of it will seep into the wax and stay there. Bees bring to the hive agricultural toxins that also accumulate. Removing contaminated comb is the most effective thing you can do to take care of the bees that take care of the bees that go into winter. When you can't see sunlight through a comb with wax foundation, it's time to replace.

From the Past

If you want to explore good wintering techniques I strongly recommend you find and read books on beekeeping published before 1930. This was when wintering was considered a significant problem and a lot of research and thought went into the problem. (More recent books on the subject, with excellent information, are found in the Resources section, but include *Beekeeping in Western Canada* and *The Hive and the Honey Bee.*

Even earlier, L. L. Langstroth wrote brilliantly about winter in his first editions:

"If the colonies are strong in numbers and stores, have upward ventilation, easy communication from comb to comb, and water when needed, and the hive entrances are sheltered from piercing winds, they have all the conditions essential for wintering successfully in the open air."

Basic Winter Checklist

Mites and diseases under control

More than enough food:
- Honey and pollen

Food and space in the right place:
- On the sides of the cluster in the bottom box
- Above the cluster in the top box

In the top box (essential):
- Pollen and honey above the cluster in the top box
- Room in the center for bees and food and pollen, with a number of empty cells in the middle

Queen:
- Less than two years old, less than one year old is better (see chapter 3 about long- lived queens)
- Productive last season, meeting all criteria you need
- Performing as expected—none, some, lots of brood, depending on the race of bees you have and how they normally perform during winter (Italians: lots of brood; Russians: almost none; the others: somewhere in between)

Basic protection:
- Mouse guards on before mice get in
- Entrance not blocked with reducer to allow adequate ventilation
- Fenced from cattle, bears, vandals
- Exterior cracks, chinks, holes sealed
- Screened bottom boards partially closed, half or so is fine
- Set slightly tilted forward so water (rain, melted snow) runs out of the door
- Solid windbreaks installed
- Wrapped or wrapped/insulated (if necessary)
- Set to be above and protected from spring floods
- Accessible throughout winter and spring
- Cover secure (tied down, secured with hardware, weighted with a large rock)

If pollen stores are light in the fall, you can feed pollen substitute patties. These will help the adult bees going into winter gain the necessary "fat body" adults need for a long winter. The bees will not, however, store this food for later use. It won't do the spring brood any good, so don't be fooled into thinking you've solved the shortage problem in the spring by feeding these in the fall.

Take Care of the Bees

Basically, all colony management focuses on taking care of the bees. This may seem simplistic, but if you want your bees to winter well you should be working toward that goal all year long.

The most fundamental beekeeping management strategy you can adopt when preparing your bees for any kind of winter—nearly tropical to frozen solid—is this: Take care of the bees that take care of the bees that go into winter.

This is simple math. Bees that go into winter need to be 110 percent healthy. If they are damaged in any way they die sooner than they should, and the colony runs out of bees sooner than it should.

Look at it in human terms: If your grandparents have all the food they need, and have protection from pests and a clean place to live, they will take good care of your parents. Then, if your well-brought-up parents, in turn, have enough food and a clean place to live that's safe from harm, they can take good care of you. And you, well cared for, healthy, fat, and happy, can handle the worst of winters because you don't have any handicaps, scars, or worries. That's how it is with the bees.

But if your grandparents have only poor-quality food to eat, live in a rundown, drafty, and contaminated home, plus have the constant harassment of all the pests and diseases bees have, they won't be able to take good care of your parents and your parents will reach adulthood with some level of damage—to their food-producing glands, their ability to fly as fast, see as well, digest food as well, or live as long. So already your parents are working under a handicap in their ability to take care of you. Even if their life gets better, safer, and healthier after they become adults, they are already damaged and unable to take care of you as well as they could have had they been 110 percent healthy.

So, think of what kind of shape you'll be in when winter winds begin to blow. You'll be sick, sorry, and soon to be dead, even if your environment is better than your parents' or grandparents'. The damage is already done. You're a goner before you even get started. That's why I say take care of the bees (grandparents) that take care of the bees (parents) that go into winter (you).

Take care of the bees that take care of the bees that go into winter.

The best protein is pollen. Collecting pollen last summer means you'll have fresh (frozen) pollen when you need it. Mix with honey or sugar and a bit of pollen substitute to make patties, or simply pour collected pollen on an empty frame and into the cells. Place these frames above the fall cluster so when the bees move up they will be as close to the pollen as possible for brood rearing.

The Natural Question

When it comes to wintering, there is no premise of it being natural. We have so disrupted any semblance of keeping bees in a natural state that to call anything we do natural would be ludicrous. But we can work with honey bee biology. A very good example of working with natural biology is, when keeping bees in colder regions when winter clusters are routine, manipulating their honey stores in late fall so the honey is above the cluster. Storing honey below the cluster or on the sides is not working with their natural instincts, and is a lethal management choice.

In chapter 4 there were discussions about choices you can make for keeping your bees healthy. It is important to choose or develop bees that are as resistant as possible to many of the problems they will encounter.

- Use mechanical techniques and tricks to keep pest populations low.
- Make certain that the colony infrastructure is clean, stays clean, and is removed when it is no longer clean.
- Be sure there is always ample, excellent food available: in early spring, for brood rearing; in summer, for dearths anytime of the year; in autumn, for storage.
- Use feeding stimulants if needed, especially when there's little food in the colony and the bees are not eating food provided. A hungry bee is far more stressed and susceptible to many problems than a well-fed and fat bee.

Bees that go into winter do so with a huge reserve of protein-like material stored in their body. For convenience, it is simply called a fat body. It is what gets bees through the winter in good shape. They only develop it if protein is available to eat (pollen or pollen substitute) but have no place to use it (no larvae to feed). Their bodies convert the built-up protein into a storage medium for them to draw on all winter long and into the spring. So, late summer is a good time for feeding the bees protein. You can measure the amount of pollen to see if they have enough and if not, feed, feed, feed. Keep feeding, in fact, until the weather won't let you get into the colony, or they absolutely won't eat another bite.

The New Nosema

At the time of this writing, the new nosema has not sorted itself out. It is closely associated with viruses and pesticides in causing honey bee deaths and reducing life spans. Even the spectre of colony collapse disorder is mentioned in the same breath as *Nosema cerenae* when coupled with viruses or pesticide exposure. It's known that bees with nosema, when fed well and continuously, using feeding stimulants if necessary, are better able to handle infestations than bees that do not eat or do not eat enough. The only other recourse is the antibiotic that reduces colony outbreaks by slowing spore germination in the honey bee's gut. It's a drug. It helps.

There are options that reduce the severity of the infestations, and other options that reduce the severity of the symptoms. The research is ongoing and any answers we could offer here would only be short-lived as new information becomes available. But stay in touch with this silent killer. And do all you can to prevent it—removing inoculum, reducing the symptoms if they show up, and feeding lots and lots.

Infrared Winter Pictures

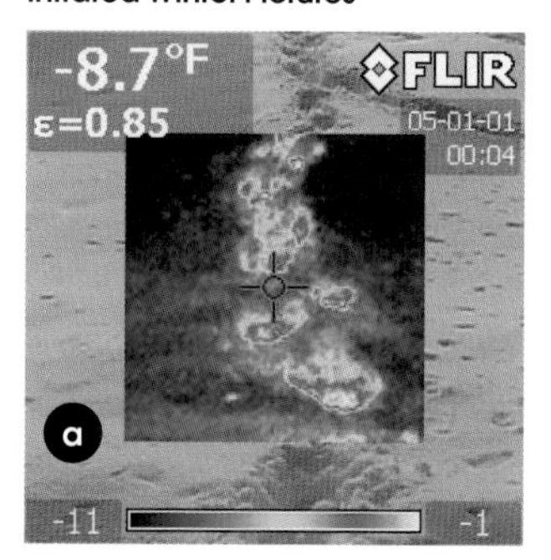

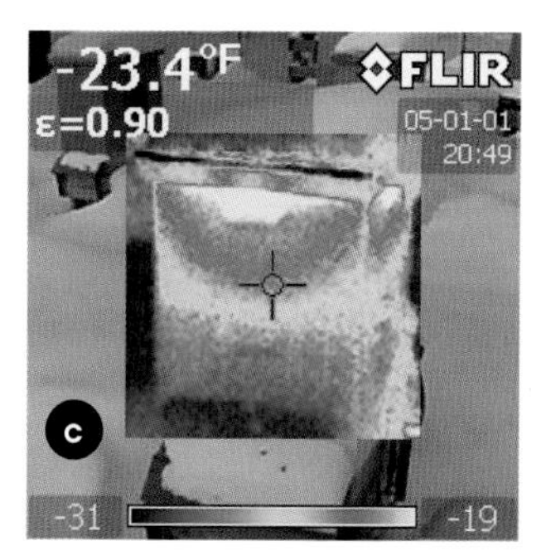

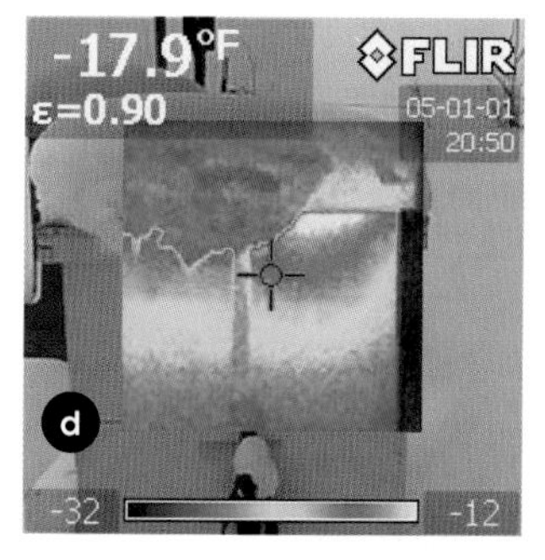

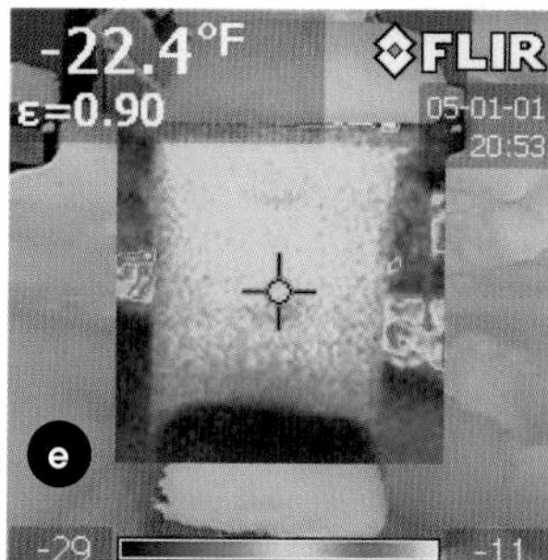

Unwrapped hives on a near 0°F (18°C) day: (a) Infrared (IR) photos of footsteps in the snow: (b) IR photo showing a large cluster—the red area occupying all of the top box and a little bit of the bottom box: (c) Cluster on the top of the box : (d) Showing snow on the roof; (e) Not a lot of heat is being generated here. This colony could be in trouble.

The Golden Rule

The golden rule of winter management, no matter where your bees are when they form clusters, is:

Enough good food in the right place.

It is that simple; but here's where it goes wrong.

Cluster Dynamics. Bees group together near the center—top to bottom and side to side—of the bottom box. You've made sure they are in the bottom box when it's time to cluster. The quantity of bees varies by race: many (Italians) or not so many (all the rest). If there are high numbers, they will be on as many as eight or nine frames (usually six or seven). They may even be on two or three frames in the top box too. But if there are not so many, they may be on only four frames in the bottom box.

Frames. The frames in the center of the cluster at the beginning of winter should have some honey and pollen, and some empty space, maybe some brood. There should be more honey on the edges of the cluster closer to the bottom than nearer the top. There should be the most pollen nearer the top, where the bees will be when they start raising brood again in the spring. A common mistake is when frames of honey and pollen are moved down so the honey is where it needs to be and the pollen goes with it, and then, come spring, there's nothing left above to feed the brood, so less brood is raised.

A frame with some honey in the corners, pollen around the brood, and lots of brood in the center will be devoid of resources in a few short days. And when the brood emerges, there will be too much empty space. Honey and pollen should be closer to the center, with just a little sealed brood so the bees can use the cells to warm new brood and be close to food.

This frame is a gold mine of reserves. It has both honey and pollen, and will do well above the cluster so they can move toward it. Do not, however, move it down early on so the cluster has the honey, because the pollen will go with it, and be useless later when brood rearing begins. Leave it above the cluster.

Bee Movement. Bees need to get into cells on both sides of frames so they have almost complete contact—on both sides by being in cells (and warming next-door brood), and over the top, underneath and around the sides of frames. This mass needs to be as much in touch as possible. The bees in the center of the ball are generally closest to the food, are in the warmest location, and so are able to do the most work. They keep busy, moving to and from food if it's close. Some care for brood and the queen if needed, and many spend a lot of time rapidly vibrating their wing muscles, which are the largest muscles in the bee's body. This exercise generates heat, which is what keeps all these bees warm. As the warm air rises, it is slowed and shifted by all the hairs on all the bees it comes in contact with. The more bees, the more the heat is held in.

Insulator Bees. Around the outside edge of the cluster is a layer, two to five bees deep, of insulator bees, with their rears in the air. The colder it is, the closer they huddle, trapping as much of that warm air in their furry layers as possible. Even colder, the layer thickens, and the bees get even closer, keeping that warm air below them, not letting it escape upward. Some insulators move into the cluster during winter; some, it is reported, stay in place all winter long. Those that move toward the center get fed by bees moving honey to them from the center. There are more insulator bees on the bottom two-thirds or so of the cluster because the warmed air tends to move up, warming the bees in the top of the cluster.

Honey. Some bees in the cluster are in direct contact with honey in the frames. Warm bees that are close to honey will enjoy it themselves as well as pass it along to those who can't reach it, in the same way foragers pass nectar to house bees when they return. A hungry bee will beg for food from her neighbor. The bees will pass the request upstream, so to speak, and the tone of the message (urgent, casual, ambivalent) is conveyed as well. The bees near honey ultimately get these messages and pass along food as it's needed. When the bees near the food have consumed it all, they will move or if they can, start begging it from others.

Appetite Control. If the weather warms up, and it is warm enough inside the box, the bees will move so more are near honey. If the weather doesn't warm up, there's no regulatory, disciplined group of bees in a colony policing the food situation. Nobody is checking out the resources, counting heads and saying, "Hey, slow down, eat less or we're going to run out of food in no time at all. Make sure the queen gets what she needs, though." If there's food available somewhere, everybody eats like there's no tomorrow. And if they eat all the honey in the hive, or can't access the honey, they all starve to death around the same time.

If you have a large population of bees they will reach far and wide. Big populations have the advantage of being able to get some of their bees way over to the side to get at that honey so it can be shared with all the bees. Bees don't move sideways unless it's warm enough and there's honey to be had over there. Moving sideways is abnormal, difficult, and often dangerous. Too much honey should not be placed on the side of a cluster.

Maintaining enough bees to reach around the frames, go over the frames, and go under the frames is one way to ensure that they will be able to reach all the food they want when they want it. These bees should be young, fat, and healthy.

How Much Honey?

My intent is to not find out how little honey I can leave on a hive and have it survive a winter. My intent, every year, is to leave so much honey on that the bees could not eat it all, even if winter lasted until the middle of summer. For a strong colony, with eight to twelve frames of bees in late fall, some brood still coming, 100 pounds (45.4 kg) isn't too much honey to be in the two brood chamber boxes, or even three boxes, as is often the case in real winter locations. Others think that is extravagant, and it's true: That's more than I've ever seen a colony eat.

There should be some honey left when you make your first spring inspection, and the bees shouldn't be at the top of the top box, whether two or three boxes—maybe two-thirds of the way up, but no more. For darker bees, with a smaller cluster and probably no brood, 80 pounds (36.3 kg) will be plenty of honey to outlast winter. That means a two-story colony will weigh 150 to 160 pounds (68 to 72.6 kg) total (equipment, bees, and food). A three-story box with extra food will weigh between 220 and 250 pounds (99.8 and 113.4 kg) total. (See the box on page 154, how to weigh a hive, and why it is important.)

A well-protected colony, with enough bees, enough food in the right place, and few pests will almost always do well. But food consumption isn't linear. The rate of consumption increases in early spring and even late winter when brood production begins. A rule of thumb is that 80 percent of the food consumed during the whole winter will be consumed in the last 20 percent of the winter. When brood starts, food goes fast and running out is not allowed.

If you are producing queens and one of the attributes you are choosing is good overwintering, food consumption is a metric you need to monitor. But you can do that with too much food in the colony, rather than not quite enough. Colony weights before and after winter will indicate consumption only roughly, because honey consumed is turned into more bees and brood. So make sure a measure of population and brood production enters into that figure. A cautious and careful colony will eat only as much as it needs, as will a large and ambitious colony. If cautious is your goal, weigh before, during, and after, and determine the most frugal colony.

The configuration for 8-frame equipment is different from 10-frame equipment. Mediums will differ from deeps, and using a mix of mediums and deeps will be different still. Figure frame surface space when comparing to 10-frame deeps for amount of honey and pollen needed. The weight of an 8-frame colony with approximately the same amount of bees, honey, and pollen will be a little less because you will not have empty frames on the outside of the supers with 8-frame equipment, as the bees utilize the space they have better. But the tower will be taller, that's for sure.

Movin' on Up

All the warm air generated by all those warm bees eventually escapes and moves upward. It's easier for the bees to follow that warm air up than to brave the cold air on the sides unless the ambient air warms enough that they can move. As food is consumed in the bottom box where they started, the next meal is just above, where the beekeeper made sure it was. And it's easy to reach because it's warm up there. So the whole colony inches upward all winter long following that warm air plume. They'll inch outward if it's warm enough and there's honey there. But going sideways can be dangerous. You can only go so far because of the walls, and once there, there's probably little stored above you because it's all in the center, where the beekeeper put it. And getting back to the center means covering a lot of ground with no honey in between because it's already gone. The bees are between a rock and a hard place.

You can see why bees starve, and why they shouldn't. There should be enough food in the hive in the fall to feed all the bees until after the spring honey and pollen flows start, whenever that is. But all that food has to be in the right place because the bees move upward, following the warm air plume all winter.

Too Cold Too Long

"Too cold too long" is often cited by beekeepers as the reason their bees died. They just couldn't move those last few inches to their next meal. That is true in an unprotected hive. The air around the bees is the same temperature as the air outside. The temperature of the wooden box sides is the same as the temperature outside. Only the bees, the comb they are covering, and any brood and food in that comb are warm. Maybe it's a bit warmer than ambient slightly above the cluster, if the air is calm and there are enough bees. If the temperature inside the hive drops to below 50°F (10°C), the bees can't move much beyond the warmer cluster. If it's very cold the warm air plume isn't enough to warm the space above the bees and they can't even move up. Doomed ...

This is as sad as it gets. Several things went wrong here. First, there weren't enough bees in the hive to be able to reach the food. The population dwindled during the winter due to damage from a variety of sources. If you look carefully, you can see the telltale fecal deposits of varroa mites in the empty cells, probably the chief cause of their early demise. You can see where the cluster was by the consumed honey. As the population dwindled they were farther and farther from their food, eventually unable to reach it at all. They simply starved to death, inches from salvation. Care earlier in the summer, and good protection in the winter, would have saved this colony.

Protection

… That is unless a barrier of some type is placed around the hive, such as some insulation or a windbreak outside to reduce the air movement inside the hive. For generations, beekeepers in the cold regions of the world have wrapped their hives in protective coverings. These could be as simple as large boxes that generously fit over the hives with the empty space filled with insulating material. Or they'd use roofing paper, plastic sheets, wrapped insulation, or other materials to cover the outside of the hives.

Good windbreaks near the hive are essential. Evergreen trees, fences, and buildings all work to block prevailing winds. But the infrastructure of the hive should be the first line of defense against the wind for the bees inside. Repair or replace all cracks, crevices, knotholes, worn-out corners, busted tops, and shoddy bottom board rails when the weather is still kind.

The air inside the hive isn't heated, but the warm air plume, if subjected to steady breezes, does not provide a haven of warmth for the bees to use to move—it is dispersed. The solution is a good windbreak and sound equipment without cracks and holes. In areas of mild winter temperatures a windbreak is usually sufficient to reduce winter stress in a hive and let the bees winter in peace. This should be the minimum protection in areas that do not go below freezing even once all winter.

Areas where the lowest temperatures dip below freezing only once or twice a winter are more stressful than the tropics but still pretty mild. To protect bees in this environment, again, start with sealing all the cracks and crevices and get a good windbreak in place. Then, protect the colony with a thin, dark, uninsulated material—a wrap of roofing paper or a carton that slips over the hive. This way, on even the coldest day, when the sun shines, the dark material absorbs the warmth of the sun and passes that into the hive. This warms the air inside and it rises the same as the warm air from the bees, but it comes from the sides so the warm air plume is wider and bigger, allowing the bees to move, maybe only a little, to their next meal.

Colonies situated with a vegetative windbreak against the prevailing wind will do better all year, but especially in the winter. These colonies face southwest and get warming sun and have protection from cold winds. In a mild climate, this is probably all the winter protection your bees will need. In a more severe climate, it's a good start.

Too Warm Means Too Much Food, Too Much Money

Mostly, warmer colonies eat less food, but you have to look at it on the food consumption per bee level. How many bees are eating that food?

Colonies that are well insulated—and well insulated is dictated by how cold it gets for how long—may not need to form a cluster at all. If the sides of the hive have plenty of insulation, the bees, in their loose cluster formation, will produce a plume of warm air that rises, and disperses to fill much of the cavity. The center of the brood area will be higher than 90°F (32°C), but the area above will be habitable and bees will be able to move up and sideways to obtain food. The queen will move with the bees, and brood production will follow as food is consumed below, but there won't be that tight cluster seen when the ambient temperature is colder than 50°F (10°C). Bees that are in this winter environment will indeed eat more food—they live longer and they raise more brood later and earlier and move more, thus needing more food to remain well fed for the same length of time. They will greet the spring with more bees, more young bees, and fewer problems than a colony that has struggled all winter, just hung on, not had food for brood, and been temperature stressed the whole time, with barely enough bees to take care of itself come spring, let alone provide support for brood. A colony that has struggled all winter may be heroic, but it is a pathetic producer the rest of the season. Why would you do that to a colony?

Providing and installing winter protection, leaving more honey to ensure adequate food for a large population until late spring, the labor to remove the material come spring, plus a place to safely store it all until next season—these all add to the cost of managing your bees. But the payback is in the spring when your colonies are more populous early on and have more than enough bees to take advantage of the really early flows. Enough bees, you'll note, to make surplus honey on the crops you normally only build bees on. Enough bees to make splits early if needed, or to raise queens early if possible, enough bees to sell early—finally, enough bees.

When it comes to preparing your bees for winter, being frugal saves a little, but costs a lot.

From the Archives: Colonies in Winter

A minimalist approach. The bees begin the winter in the bottom of this $1^{1}/_{2}$-story colony with some honey below, but most above, along with ample pollen. The colony is tipped forward slightly, a piece of roofing material is stapled over the front entrance to keep the wind out, but the top entrance through the inner cover slot is left open so the bees can exit on either side. The cover is secured with a large rock. In early spring the bees will all be in the top box, the medium below is removed, and another deep is added. By the time the bees move up, swarm season is almost over and they have had ample room to expand and grow.

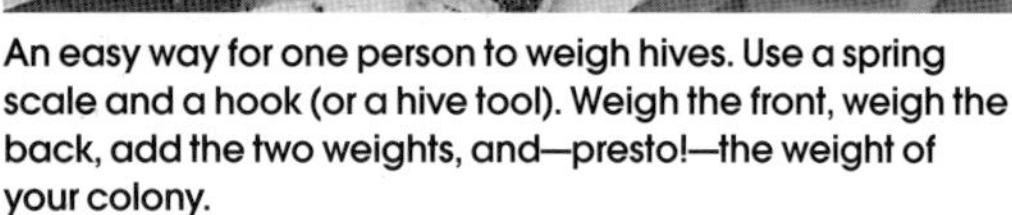

An easy way for one person to weigh hives. Use a spring scale and a hook (or a hive tool). Weigh the front, weigh the back, add the two weights, and—presto!—the weight of your colony.

Overwintering bees in cellars or underground shelters was once common. Today, beekeepers in the coldest regions approach this a bit differently. Sophisticated ventilation systems keep the rooms at the right temperature, carbon dioxide levels at the right level, and cool enough long enough so that when the bees come out in the spring the environment is ready for them to expand rapidly. Indoor wintering is costly to set up, but not too expensive to maintain, and saves a lot of bees.

Another sad story here. Extremely poor ventilation means warm, moist air can't escape. Instead, it condenses on the bottom side of the inner cover and then drips back into the colony. When it collects on the top bars it freezes but then melts when the temperature warms and drips back on the bees, resulting in soggy bees. Note, too, the fecal-stained entrance, probably the result of *Nosema apis*, which hasn't been a problem for several years.

Wrapping colonies (here, two) with roofing paper was common before colonies were palletized. This still works for climates that aren't too harsh with single or more colonies at a time. Storing the wrapping material, however, can be troublesome.

Medium Wrapping Ideas

This level of protection will be considered overkill by many beekeepers. But I'm not in this to see how little I can spend or to find out how tough my bees are. I'm in this to pamper and protect and promote the health and welfare of my bees. You've heard it before: Take care of your bees and they'll take care of you. Taking care of bees in winter is an old problem. We can't worry about newer problems and let the old ones slide.

Colonies in very cold regions are wrapped, usually in fours on a pallet, but occasionally in rows, with insulation, and then covered with a layer of insulation. Then the cluster is covered in plastic, with a common roof, usually of plywood. There is no bottom entrance. Colonies treated like this will do quite well and be able to move much of the winter. Obviously, uncovering to emergency feed is not recommended.

Use materials that are flexible, with a thick plastic shell and an insulated core. The black color heats well even on very cold days, warming the inside of the colony, and the insulation extends the time the inside is warmed. Durable and easy to store.

Collapsible corrugated cardboard or plastic boxes work well for single colonies, even in harsh climates. Insulation can be placed in the space, especially if you are using 8-frame equipment, or the space can remain empty. The black color warms the space and inside the hive very effectively. Storage can be an issue, and if you have lots of colonies this might not be the answer.

Ventilation

Every beekeeper knows it's not the cold that kills the bees, it's the wet that kills them (or they starved, or they got dysentery). Too often secondary causes occur, but wet is a real killer. As that warm air plume rises it eventually reaches the top of the colony, usually the bottom of the inner cover or the migratory cover. That piece of wood is the same temperature as the outside ambient air. Warm air, cold surface, condensation. Accumulated condensation—cold water—has nowhere to go. It drips and drops and falls back into the hive. If it gets cold enough long enough, and there's enough water forming on the bottom side of the cover, what happens is that water will drip down into the colony. Sometimes it will collect on the surface of the frames in the top box, and on the top bars of the box below the top box and freeze there, effectively closing off the top box of honey to the bees. What you find in the spring is a lot of dead bees a box away from the top box that's full of honey and you wonder what went wrong because by then the ice is long gone.

To allow warm moist air to escape rather than condense inside, an upper entrance should be provided. An inner cover with an escape hole, with a notch cut out of one side of the rim that goes around the edge, acts as the top entrance. But above the inner cover is a slab of foam insulation. The hot air, and the bees, can get out of the hive by going through the hole in the edge of the inner cover that's on the super side of the cover. The inner cover stays warm enough because of the warm air rising and the insulation on top so that the moisture doesn't condense but rather flows on outside. If the inner cover is flat, that is, doesn't have a rim, a groove can be carved in the insulation beginning with the escape hole in the center of the inner cover that leads to the outside. That way, the bees can leave through the inner cover escape hole opening and get outside by following the groove in the insulating board. If the inner cover has a rim without an upper entrance notch, and many do, cut the insulation to fit snugly inside the rim flat against the inner cover surface and cut a notch in the inner cover's rim to correspond to the notch in the sheet of insulation for moist air and bees to use if needed.

Some beekeepers use an extra super, placed above the top super the bees use, and fill it with leaves, newspaper, you name it. The moisture travels through this super to escape. An inner cover is between them and the notched entrance is available from the super below. The escape hole is screened to keep bees out. Warm air rises through the hole, hits the cold, condenses, and is captured in the absorbent material and stays there. There are pieces of equipment you can purchase that collect condensation and channel it out through tubes or spouts.

There are a hundred ways to insulate, absorb, and ventilate. The technique used is less important than making sure there is an upper entrance for the bees if the bottom entrance (if there is one) gets clogged, and a way to move moist, wet air out of the way. The simpler the technique, the better, though, especially if you are wrapping your colonies.

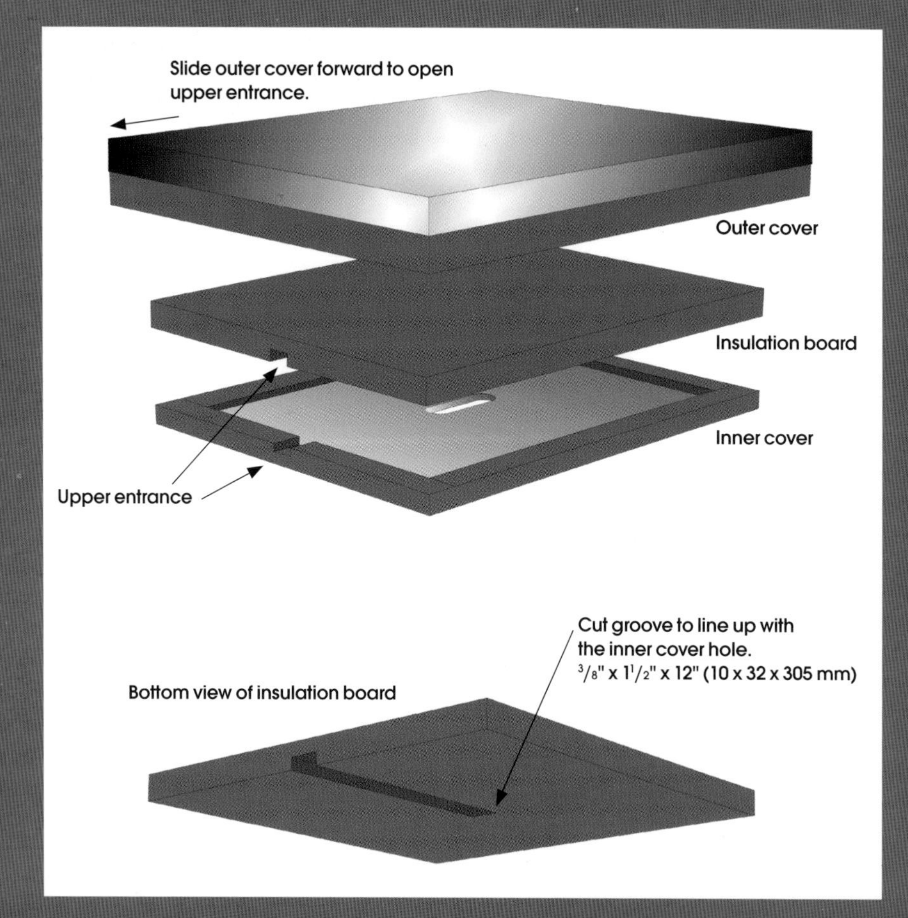

Placement of Grooved Insulation Board

When It's Really Cold

In severe winter locations a well-insulated colony:

- Eats more (probably).
- Is buffered from harsh and rapid outside temperature changes so the bees aren't caught spread out and unprepared.
- Keeps out any hint of drafts or wind, which keeps the whole inside of the colony warmer.
- Is drier, especially on the sides of the boxes.

These long, cold, wintering locations require hardcore insulation, wrapping with weatherproof material, excellent ventilation through a top entrance, and a domed top for snow removal. Colonies are wrapped in groups and the efficiency of wrapping comes into play here. One at a time is costly in labor and material. Four seems about right, so that two sides of a colony are kept warmer than the two sides exposed, as it were, to the insulated outside. Rows accomplish the same thing—two sides warmed, two exposed. Some feel that if three or all four sides are adjacent to another colony the interior would actually be too warm, the bees would eat too much, raise too much brood, and cost more to keep. And, the good beekeepers I know who have experimented with this sort of intensive wrapping will also tell you that when ganging colonies in four by six groups wintering works all right, but working the colonies in the center of the group is difficult.

I don't know the answer to the too-warm analysis, but I suspect the food consumption part is correct (warmer equals more brood and more bees). I don't know if the payback is great enough to compensate in areas that are really cold, and have long, long winters, and late, late springs. If you try it, let me know how it works.

Planning Your Wrap

When you plan your wrap, consider that you may have to take it apart in early spring to feed. Can you do that and put it back together in a minimum amount of time without damaging the material and interfering with your colonies? If you build it, can you build it again? Make sure you can, for all the times you screw up and take too much honey and underfeed in the fall.

All the rest of the typical winter warnings hold in really cold places. Keep colonies away from areas prone to spring floods, provide more than enough food in the right place, and even insulated colonies are aided by having wind blocks of some kind. The ventilation aspect does change though. A bottom entrance may not even be provided because it will be partially to completely blocked by snow and ice for much of the season (obviously, screened bottom boards are partially to mostly, but not completely, covered). Only a top entrance is provided, which also serves as the only ventilation port. But a colony that's warmer inside will have less condensation and so will be drier anyway.

Insulation should be applied after autumn has set in so the colony has finished maneuvering its food and brood, and should be left on until spring weather is settled. The bees do have a top entrance after all, and even if they are in the top box come spring, if you provided enough honey and pollen they will move down to store early food and raise brood. What do you think they do in trees?

If only we could depend on the world to supply all these things in the right amounts at the right time ... so we could have the right number of bees in the right place at the right time. But the world doesn't. So we must.

Pollen

There needs to be pollen stored in your overwintered colonies. This provides the nutrients necessary to raise brood, throughout the winter for some colonies but certainly in late winter and early spring for all colonies. Without adequate pollen, quantity and quality, the colony will dwindle or collapse. If old bees are not replaced by younger bees raised during the winter, the colony follows suit.

Like honey, pollen needs to be in the right location, too. It all needs to be close to the center of the cluster space but not choking it, and more needs to be in the second or third box than in the bottom or second box (though some needs to be in the bottom).

More than eighty years ago researchers from Wisconsin recommended at least 500 square inches (0.3 square m) of pollen should be in an overwintered hive by late fall in moderate to severe winter locations.

A deep frame has, counting both sides, about 275 square inches (0.2 square m) of cell space. Two frames, completely full of pollen, then would be sufficient. A medium frame has 170 square inches (0.1 square m), so three medium frames, completely full would hold enough pollen.

But you know that pollen, generally, isn't placed in frames side to side and top to bottom. A more general guideline is several deep combs and several more medium frames, all partially filled. Recall the exercise in estimating brood on a frame (see the section on swarming). The same math applies here relative to estimating the number of cells on a frame. Once you get a handle on how many cells, whether brood or pollen, recall that there are 25 cells per square inch (6.5 square cm). So a minimum of 500 square inches (0.3 square m) is nearly 13,000 cells of pollen that need to be stored before winter sets in. Do the math before it's too late so you can share pollen from a pollen-rich colony, or be prepared to feed a pollen substitute or supplement early in the spring.

There is, of course, the pollen that's stored under honey. That changes the equation, but my opinion is that if you can't readily tell if there's pollen under the honey, assume it isn't and plan accordingly. Having too much is safer than running out early.

500 square inches (0.3 square m) of stored pollen to make it through the winter.

Common Sense and Practical Facts

There's an old saying (which some beekeeping researchers swear is true): It takes a cell of pollen, a cell of honey, and a cell of water to make a bee. So every bee that is raised after you shut a colony down needs those resources to grow and be strong. The water is a given. There's usually enough in a hive during the winter (just make sure not too much). And with 100 pounds (45.4 kg) of honey in your colony you'll easily have enough of that resource. So any limiting factor to brood production will be the amount of pollen available. If you have 13,000 bees—weighing roughly 4 pounds (1.8 kg), raised between late fall and spring—how much honey will just those bees need? And when will they arrive and begin eating it?

Getting into a colony in the subtropics during your downtime to measure sealed brood and calculate the egg-laying rate of your queen is easy. But it's not easy in the snowbelt. It's a guess, actually. It's wise to overestimate the amount of honey the colony will need. And make sure you know how much pollen the colony has. A shortage of either resource will be a limiting factor in the amount of brood the colony can raise.

The penalty for not having enough pollen stored in the colony when cluster time begins is that you will have to feed pollen (or some kind of protein) in late winter or very early spring to compensate. A problem ensues: A colony that's wrapped will need to be unwrapped and entered during what can only be less-than-desirable conditions for opening a colony. If you don't feed protein, the colony dwindles and either dies or is so underpopulated when real pollen becomes available, that it is worthless.

The goal, then, is to make sure the colony has enough pollen—good pollen—in the fall before it gets wrapped. Frames of corn pollen, for instance, are nearly worthless, though better than no pollen at all, because of its low nutrient content. But feeding pollen substitute or supplement in the fall isn't the solution. Fall fed supplement or substitute cannot be moved from a patty above to a cell below to be stored. Bees simply can't pick it up and move it. Besides, feeding pollen or protein in the fall, especially if you are feeding syrup, may initiate brood production, and that's not (usually) in the game plan for this time of year. It is beneficial to at least do a sample feeding to see if the bees are hungry. If the adults eat a protein patty in the fall they in all likelihood needed it and will convert it into that critical fat body so necessary for bees to overwinter. And, in all likelihood, they need more—right now! But they won't store it.

The exception: Bees have been known to collect and store pollen substitute when they can gather it out in the field as a powder during the hectic, it's finally warm enough spring brood-rearing time. When it is placed in a protected place to avoid rain, dew, or snow so it stays dry, foragers will dive in and cover themselves with the dust and bring it home, just like pollen (or birdseed or sawdust you sometimes see bees collect in the spring). Commercial beekeepers, strapped for time and labor, will, if the weather cooperates, simply dump the stuff right on the ground and let their bees "graze" until it's gone. No sugar is added and if the correct amount is applied after dew in the morning it will be gone before dew in the evening. This dust will be returned to the hive, dumped into an empty cell, and packed in, just like pollen. It usually doesn't last long, though, because this is going on during intense brood rearing. But in the fall, if the bees need stored protein, it is a technique to consider.

In the 1920s, a technique used to generate frames of pollen was to remove queens from a few strong colonies during the honey flow in midsummer. These colonies would continue collecting pollen for some time, but because it wasn't being used it was all stored. By the end of the flow there would be many frames of stored pollen to harvest and share with pollen-poor colonies in the fall or the following spring. The bees from these colonies would be joined to queenright colonies to beef up their populations for late honey flows, or left to perish—poor reward for the protein gold they produced. But they are old bees and not productive in an overwintering sense. And, I suspect, a clever beekeeper somewhere will find a way to incorporate this into a summer split program and get twice the reward for the effort. Think about it.

Simply moving frames of pollen from colonies with lots to colonies with little in late fall is possible. This is probably the only way to ensure adequate food without exposing your bees to the rigors of late winter exposure when feeding in a snowstorm. A metric to use as often as necessary is to offer colonies protein patties any time of the year you think there might be protein stress in a colony, including early spring, summer droughts, and fall storage time. See what the colony does in response. If they devour it and consume it before you can get out of the beeyard, you will see the status of that colony. If it is still uneaten the following week, you can relax a bit.

Summary

At the beginning of this book I talked about that meeting long ago with Dr. Shimanuki and his Rule of Rights: the right number of bees, of the right age, at the right time, in the right place, and in the right condition. "Figure that out, and you'll do just fine," he said. "It's the best set of rules you'll ever choose to follow."

To get Shim's rules working for you, you must first have bees that are healthy and living in a place that's unquestionably pristine. There must be access to abundant and excellent food year-round, and they need assistance with their greatest stress: winter. These same rules were in place when Shim spoke, but beekeepers paid little attention. The rules were accomplished with little effort or expense.

But the decades-long transition from Shim's bees to our bees has been difficult and expensive. The myopic, selfish decisions we made for our bees have polluted our small universe and contributed to the slaughter of millions of colonies. Our technology has been our undoing. Moreover, we have intentionally and with malice depleted the diversity of the diet of all pollinators to the point of ruin. As Pogo said, "We have met the enemy, and the enemy is us."

Now, we look at things differently. To grow your business, you first have to clean up your place, put good food on the table, and start making life better for your bees. We have for so long ignored these basics that we have pretty much forgotten them ... if we ever knew them at all. From today onward, everything we do must exceed the fundamentals, regardless of the cost or time it takes. There are no shortcuts to this better place.

And in growing your business, these principles must always be at the forefront. When measuring what you have, you must discount those leftovers that promote the old ways. The value of polluted equipment, especially the wax, is not the same as it was yesterday. As beekeeping equipment, its value is essentially zero.

To review, here are the key lessons I think are most important in this book:

Provide (or create) better food. Is it feasible to lease a large tract of land, get it planted with bee-friendly food, establish semi-permanent headquarters, plus build a huge apiary that supports honey production, queen rearing, and nuc production? Yes. Can it be profitable? In some places, with some landowners, yes. Everywhere? Probably not. Is it the best choice for every beekeeper? No. It would be less desirable if it were perfect for every beekeeper: We'd be awash in honey and starving for enough bees to pollinate the crops that need bees.

Beekeepers that follow the traditional model will always be needed even more. But where land is available and beekeepers are willing to pursue this adventure, the opportunities and possibilities are awesome.

If you are in a location where having long-term access to a large tract of land isn't possible or affordable you can still have a positive effect on increasing pollinator forage by planting the right crops on land that is otherwise unused. Plus, growing government programs make it possible to afford all sorts of honey bee crop improvement. In the several months since chapter 2 was written, additional funding has become available to landowners to provide pollinator forage improvement. These programs, and beekeepers asking the right questions, are already making a difference.

Raise your own queens. Raise enough to supply you and everybody within shouting distance. Raise queens that produce bees that laugh at the bad guys. Raise them any way you want. It's that simple.

Overdo winter protection. Wherever you live or whatever kind of winter you have, you must provide enough good food in the right location in your hives with plenty of time for everything to be in place before it needs to be in place (spring). Make doubly sure you take care of the bees that take care of the bees that go into winter. And then double check what you did, how you did it, and how well it worked. Take nothing for granted. Be on your toes when taking the best care of your bees that you can.

For most people, having bees is only one of the things they are doing now. But many want to make bees a more prominent part of their lives. Finding a balance between having more bees and keeping up with the day to day is the difficult part. I can't find that balance for you, but I hope some of the information you've found here makes better use of your time, while not costing you a lot of money.

To be successful at any scale—backyard, part time, or even huge and commercial—you must have a clean beehouse, enough good food, and the best bees in the world.

And you know how to do that.

So keep your veil tight, your hive tool handy, and your smoker lit. It all starts now.

Glossary

A. I. Root
Founder of the first and largest beekeeping equipment manufacturing company in the United States, located in Medina, Ohio, USA.

American foulbrood (AFB)
A brood disease of honey bees caused by the spore-forming bacterium Paenibacillus (formerly Bacillus) larvae.

annual crop
A crop that is planted, grows, is harvested, and dies in one growing season. Buckwheat is an annual crop.

Apis mellifera
The genus and species of the honey bee found in the United States.

bait hive
An empty box with an attractant inside to attract swarms.

balling the queen
When hive bees surround and kill their queen; usually a new queen.

bee bread
A mixture of fermented pollen and honey used as food by the bees.

beeswax
A complex mixture of organic compounds secreted by eight glands on the ventral side of the worker bee's abdomen; used for molding six-sided cells into comb. Its melting point is from 144°F (62°C) to 147°F (64°C).

brood
The term used for all immature stages of bees: eggs, larvae, and pupae.

brood chamber
The part of the hive in which the brood is reared.

capped brood
Pupae whose cells have been sealed as a cover during their nonfeeding pupal period.

chilled brood
Developing bee brood that has died from exposure to cold.

Cloake board
Essentially, this is a starter/finisher colony that is only one colony. The queenless unit, used as a starter, is above the queenright unit, and the two are separated by a queen excluder and a separating sheet that effectively removes any influence of the queen below. After the necessary thirty-six hours, the sheet is removed, but the queen excluder remains.

cleansing flight
A quick, short flight bees take after confinement to void feces.

cluster
When the temperature in the brood nest falls below about 85°F (29°C), house bees and workers cover the brood to keep the brood at the correct temperature. When the temperature in the hive falls below 50°F (10°C) or so, the bees gather together to keep each other warm. They form a sphere near the center of the nest, covering the brood, and forming an insulating layer of bees on the outside. Bees inside vibrate muscles to generate heat to keep the mass of bees—the cluster—warm.

colony collapse disorder
A combination of several factors, including virus infections transmitted by varroa, diminished resistance caused by nutritional and other issues, and disease, primarily *nosema ceranae*, that result in adult bees dying prematurely at a great rate, leaving in short order only the queen, sealed brood, and very young workers.

comb foundation
A commercially made sheet of plastic beeswax with the cell bases of worker or drone cells embossed on both sides.

contracts
Legal documents that spell out the rights and responsibilities of two (or more) parties in an agreement. For beekeeping, contracts are most common for pollination jobs, where the contract deals with the number and strength of colonies and time of delivery and pickup for the beekeeper, and provide safety and payment for the grower. Contracts are also used when a beekeeper leases land from a landowner. Here, the contract spells out any of many variables—uses allowed, length of the contract, how payment is made, and the like.

dearth
A time when nectar or pollen or both are not available.

dividing
Reducing a strong colony to many smaller colonies that will take some time to build back to the original size, and requeening each; or, subtracting just enough brood and adults from a strong colony to convince it not to swarm; or, splitting a large colony into essentially two equal-size colonies, such that both will be productive in that same season.

drawn comb
A comb with cells built out by bees from foundation.

drifting
When bees go into a colony that is not their own.

emergency queens
These are queens produced by a colony when the attending queen is suddenly lost, usually through an accident or disease. As soon as the colony is aware of the loss of the queen, workers select larvae to raise as a queen. Since the age of the larvae chosen is critical (twenty-four hours or less), and it is seldom known how old the larvae are that are chosen, emergency queens are always a gamble. They may be superior queens, or they may be duds.

European foulbrood (EFB)
An infectious brood disease of honey bees caused by the bacterium *Melissococcus* (formally *Streptococcus*) *pluton*.

forager
Worker bees that work (forage) outside the hive, collecting nectar, pollen, water, and propolis.

formic acid
An organic acid used to control varroa. It comes in the form of a disposable cloth that can be applied to a colony at any time of the year without fear of harming the honey or the bees.

GMOs
Genetically modified organisms. For beekeepers, GMOs are crops their bees occasionally forage on, such as a GMO canola, or GMO soybeans, both modified to tolerate herbicide applications that kill weeds, but do not harm the crop plant.

grafting
In beekeeping, grafting is moving an egg from the cell it was laid in by the queen to an artificial cell that will be placed in a starter colony on its way to becoming a queen.

guard bees
After bees have been house bees, but before they become foragers, many spend time as guard bees: stationed at the front door or other entrance, checking incoming bees to make sure that they belong to their hive.

HFCS
High-fructose corn syrup. A sweetener made from corn often fed to bees. Research has shown that feeding HFCS to bees shortens their lives when compared to sucrose as a food. When HFCS is overheated, breakdown products that are formed are toxic to bees.

honey
A sweet material produced by bees from the nectar of flowers, composed of glucose and fructose sugars dissolved in about 18 percent water; contains small amounts of sucrose, mineral matter, vitamins, proteins, and enzymes.

honey flow
A time when nectar is available and bees make and store honey.

honey stomach
A portion of the digestive system of the adult honey bee used for carrying nectar, honey, or water.

Hymenoptera
The order of insects that all bees belong to, as do ants, wasps, and sawflies.

Langstroth hive
Our modern-day, man-made, movable frame hive named for the original designer.

laying worker
A worker bee that lays drone eggs, usually in colonies that are queenless.

marked queen
A queen that has had a drop of paint applied to the top of her thorax to identify her and make her easier to find.

mating flight
The flight made by a virgin queen when she mates in the air with several drones.

mating yard
An apiary devoted to producing a maximum number of drones. Drone-laying queens, drone comb, and drone-laying workers are all used in these yards.

metamorphosis
The four stages (egg, larva, pupa, adult) through which a bee passes during its life.

migratory beekeeping
The business of moving honey bee colonies by truck to various locations to pollinate a variety and succession of crops, or to harvest honey.

monoculture
Within a given area that is large enough that a colony of bees cannot forage outside its boundries, only a single crop grows (e.g., a large corn field, soybean field, or almond orchard). Only one food source is available.

nectar
A sweet liquid secreted by the nectaries of plants to attract insects.

net worth
The difference between what you have and what you owe.

nosema ceranae
A more toxic cousin of the familiar *nosema apis*, this disease builds and is destructive more in the summer than winter, and when coupled with many viruses, quickly kills its host, and often entire colonies. Is often associated with colony collapse disorder.

nuc or nucleus (plural, nuclei)
A small two- to five-frame hive used primarily for starting new colonies.

nurse bees
Young bees, three to ten days old, that feed and take care of developing brood.

PDB (paradichlorobenzene)
Crystals used as a last-resort fumigant to protect stored drawn combs against wax moth.

perennials
Plants that, once established, continue to grow every season. Usually considered forbs or shrubs, they are long-lived but not permanent-lived plants. Wild sunflowers or honeysuckles are considered perennials.

permanent crops
These are generally considered trees. Very long-lived and grow to extreme size.

pesticides
The general name for chemicals used to kill pests of many varieties. Subcategories of pesticides are insecticides (which kill bees) and fungicides (which kill fungi, but can also be detrimental to honey bees). Combinations of insecticides and fungicides can be extremely deadly to foraging honey bees.

pheromone
A chemical secreted by one bee that stimulates behavior in another bee. The best known bee pheromone is queen substance, secreted by queens, which regulates many behaviors in the hive.

pollen
The male reproductive cells produced by flowers and used by honey bees as their source of protein.

pollen basket
A flattened area on the outer surfaces of a worker bee's hind legs with curved spines used for carrying pollen or propolis to the hive.

pollen trap
A mechanical device used to remove pollen loads from the pollen baskets of returning bees.

pollination
The transfer of pollen from the anthers to the stigma of flowers.

pupa
The third stage in the metamorphosis of the honey bee, during which the larva goes from grub to adult.

queen
A fully developed female bee capable of reproduction and pheromone production; larger than worker bees.

queen cage
A small cage in which a queen, with or without worker bees, is placed for shipping and/or introduction into a colony.

queen cell
The cell queen honey bees are raised in. Larger than worker or drone cells, when artificially produced they are often raised in great quantities to requeen colonies. Naturally occurring queen cells can be found almost anywhere in a colony, and may be the precursor to a colony swarming, or the result of an emergency queen replacement.

queen cell cup
A round, cup-based structure that workers build on the bottoms of frames to accommodate a future queen cell. The current queen must place an egg in the cup before workers begin building the rest of the queen cell. Queen cell cups are built most often just before swarm behaviors begin.

queenright
A colony with a healthy queen.

robbing
When bees steal honey, especially during a dearth, generally from weaker colonies.

royal jelly
A highly nutritious glandular secretion of young bees, used to feed the queen and young brood.

Russian bees
a line of honey bees that had spent generations exposed to varroa mites without miticides. They were brought to the United States from eastern Russia for their innate resistance to mites.

scout bees
Foraging bees, primarily searching for pollen, nectar, propolis, water, or a new home.

small hive beetle
A destructive beetle that is a beehive/honey house pest, living generally in the warmer areas of the United States; originally from South Africa.

spermatheca
An internal organ of the queen that stores the sperm of the drone.

starter/finisher
One method to raise queens using two different colonies. Grafted cell cups are placed in queenless colonies with many young nurse bees. Queenless colonies are more inclined to accept eggs and begin queen cells because they are queenless; these are the starters. After about thirty-six hours, the now-started queen cells are moved to a queenright colony, the finisher, which is more inclined to complete the process capping the queen cell because it is imitating a swarm-inclined colony. Between the two, queenless and queenright, the grafted queen cells receive excellent care.

supersedure
A natural or an emergency replacement of an established queen by a daughter in the same hive.

surplus honey
The honey stored by bees in the hive that can be used by the beekeeper; it is not needed by the bees.

survivor queens
Queens produced by queens that lead colonies that thrive in the face of disease and pest exposure are considered survivor queens. The bees in these colonies require little or no treatment for these pests, and seem to thrive in spite of them. They are produced without any focus on breeding or selection; if they are alive, they are survivors.

swarm
When about half the workers, a few drones, and usually the queen leave the parent colony to establish a new colony.

swarm cell
A queen cell especially produced by workers when a colony is nearly ready to swarm. Eggs are placed in special queen cups, usually but not always toward the bottom of frames, and many, many are raised. When the cell is capped, the swarm issues.

Tylosin
One of several antibiotics used to treat, but not cure, American foulbrood.

uniting
Combining two or more colonies to form a larger colony.

varroa mite
Varroa destructor, a parasitic mite of adult and pupal stages of honey bees.

virgin queen
An unmated queen.

VSH (Varroa Sensitive Hygiene)
Varroa Sensitive Hygiene honey bees can locate a larva in a capped cell that contains varroa. VSH bees remove the infected larva and mites, and thus reduce varroa populations in a honey bee colony.

worker bee
A female bee whose reproductive organs are undeveloped. Worker bees do all the work in the colony except for laying fertile eggs.

worker comb
The comb measuring about five cells to the inch in which workers are reared.

Resources

BOOKS

The ABC & XYZ of Beekeeping, **41st Edition***
Ed. by H. Shimanuki, Ann Harman
A. I. Root Co.

Africanized Honey Bees in the Americas
Dewey Caron
A. I. Root Co.

*The Backyard Beekeeper's Honey Handbook**
Kim Flottum
Quarry Books

The Bee Book Beekeeping in the Warmer Areas of Australia
Warhust & Goebel
DPI, Queensland

Beeconomy: What Women and Bees Can Teach Us about Local Trade and the Global Market
Tammy Horn

The Beekeeper's Garden
Hooper & Taylor
A & C Black Plc, London

The Beekeeper's Handbook
Sammataro & Avitabile
Cornell University Press

The Beekeeper's Lament
Hannah Nordhaus
HarperCollins

Beekeeping for Dummies
Howland Blackiston
Hungry Minds Press

Beekeeping in Western Canada
Ed. by John Gruszka
Alberta Agriculture

Bees and Beekeeping
Eva Crane
Comstock Press

Bee Sex Essentials
Lawrence Connor
Wicwas Press

Bees in America
Tammy Horn

*Breeding Super Bees**
Steve Tabor
Root/Northern Bee Books

*The Buzz about Bees**
Jürgen Tautz, H. R. Heilmann, and D. C. Sandeman
Springer

Control of Varroa for New Zealand Beekeepers
Goodwin and van Eaton
New Zealand Ministry of Agriculture

*Dance Language of the Bees**
Karl von Frisch
Harvard University Press

An Eyewitness Account of Early American Beekeeping
A. I. Root Co.

*Fat Bees, Skinny Bees**
Doug Somerville
Rural Industries Research and Development Corporation

Following the Bloom
Douglas Whynott
GP Putnam

Form and Function in the Honey Bee
Lesley Goodman
IBRA

Fruitless Fall
Rowan Jacobsen
Bloomsbury USA

The Hive and the Honey Bee
Ed. by Joe Graham
Dadant and Sons

***The Hive and the Honey Bee*, 3rd Edition, Annotated**
Roger Hoopingarner
Wicwas Press

The Honey Bee Democracy*
Thomas Seeley
Princeton University Press

The Honey Bee Inside Out
Celia F. Davis
Bee Craft Ltd.

Honeybee: Lessons from an Accidental Beekeeper
C. Marina Marchese
Black Dog & Lenenthal

Honey Bee Pests, Predators, and Diseases
Ed. by Morse and Flottum
A.I. Root Co.

Honey, the Gourmet Medicine
Joe Traynor
Kolvak Books

Honey Plants of North America*
(reprint of 1926 edition)
Harvey Lovell
A. I. Root Co.

The Honey Spinner
Grace Pundyk
Pier 9

Increase Essentials*
Lawrence John Connor
Wicwas Press

Keeping Healthy Honey Bees
Aston & Bucknell
Northern Bee Books

***Langstroth's Hive and the Honey Bee* (reprint of 4th Edition)**
L.L. Langstroth
Dover Books

Natural Beekeeping
Ross Conrad
Chelsea Green Publishing

Plants and Honey Bees*
David Aston and Sally Bucknall
Northern Bee Books

Pollen
Kesseler & Harley
Papadakis

Queen Bees
David Woodward
Northern Bee Books

Queen Rearing and Bee Breeding
Laidlaw and Page
Wicwas Press

Queen Rearing Essentials*
Lawrence Connor
Wicwas Press

The Quest for the Perfect Hive
Gene Kritsky
Oxford University Press

Removing Bees*
Cindy Bee and Bill Owens
A. I. Root Co.

The Superorganism
Bert Hölldobler and Edward Wilson
W.W. Norton & Co.

Sweet Journey: The Biography of Nephi E. Miller
Miller Family Trust

Weeds of the Northeast*
Uva, Neal, and Thomas
Cornell University Press

Weeds of the West*
Ed. by University of Wyoming
Western Society of Weed Science

What Do You Know?
Clarence Collison
A. I. Root Co.

Where Have All the Flowers Gone?*
Charles Flower
Papadakis Publishers

The Wisdom of the Hive
Thomas D. Seeley
Harvard University Press

The World History of Beekeeping and Honey Hunting
Eva Crane
IBRA

A World Without Bees
Alison Benjamin and
Brian McCallum
Guardian Books

*Recommended

MAGAZINES

North America

American Bee Journal
Dadant and Sons
www.dadant.com

Bee *Culture, The Magazine of American Beekeeping*
A. I. Root Co.
www.BeeCulture.com

Hive Lights
Canadian Honey Council
www.honeycouncil.ca

Australia and New Zealand

Australian Bee Journal
abjeditors@yahoo.com

The Australasian Beekeeper
www.penders.net.au

The New Zealand Beekeeper
www.nba.org.nz

Europe

An Beachaire
The Irish Beekeeper
www.irishbeekeeping.ie

BeeCraft
British Beekeeper's Association
www.bee-craft.com

The Beekeeper's Quarterly
www.beedata.com

Beekeeping
www.devonbeekeepers.co.uk

Bee World
Journal of Apicultural Research
www.IBRA.org.uk

La Sante de l'Abeille
www.sante-de-labeille.com

The Scottish Beekeeper
www.scottishbeekeeper.com

Vida Apicola
www.viaaapicola.com

Suppliers

B & B Honey Farm
Houston, Minnesota, USA
www.bbhoneyfarm.com

BeeOlogy
www.beeology.com

Brushy Mountain Bee Supply
Moravian Falls,
North Carolina, USA
www.beeequipment.com

Dadant and Sons
Hamilton, Illinois, USA
www.dadant.com

Dakota Gunness, Inc.
dakgunn@rrt.net

Glory Bee Beekeeping
Eugene, Oregon
www.glorybeefoods.com

Walter T. Kelley
Clarkson, Kentucky
www.kelleybees.com

Mann Lake, Ltd.
Hackensack, Minnesota, USA
www.mannlakeltd.com

Maxant Industries
Ayre, Massachusetts, USA
www.maxantindustries.com

Penders Beekeeping Supplies
Australia
www.penders.net.au

Pierco Frames
Chino, California, USA
www.pierco.net

Ross Rounds
Albany, New York, USA
www.rossrounds.com

Rossman Apiaries
Moultrie, Georgia, USA
www.GaBees.com

E. H. Thorne
Wragby, UK
www.thorne.co.uk

Index

Photographer Credits

All photography by Kim Flottum with the exception of the following:

Courtesy of A. I. Root 53 (right)

Yeshwant Chitalkar/ Red Hook Honey 23 (top)

Fotolia.com 12; 13

Cindy Hodges 29

© FLPA/David Hosking/ agefotostock 65 (bottom)

iStockphoto.com 48; 49 (right); 56; 57 (right); 60; 65 (top); 66 (top, left & right, bottom right); 67, (top right, middle, & bottom left); 68; 69 (top, right); 72 (top, left & bottom); 74; 114

Dennis LaMonica 147 (top row & bottom, right)

©Tom Mackie/ alamy.com 45 (top)

Courtesy of Maxant Industries 17 (bottom)

Gwen Rosenberg 23 (bottom)

Shutterstock.com 5; 7; 15 (bottom); 32 (bottom); 52 (left)

Courtesy of the United Sates Department of Agriculture 8; 81

©Leonard F. Wilcox/ alamy.com 50

Paul Younger 47 (right)

About the Author

Kim Flottum has been at the helm of *Bee Culture* magazine for just over 25 years. He is the author of *The Backyard Beekeeper* and *The Backyard Beekper's Honey Handbook*, both published by Quarry Books. He keeps bees, gardens, and writes for a host of publications about both of these subjects, but mostly about bees and beekeeping.

Acknowledgments

Special thanks to all the folks at Quarry that combine their talents to make these books read so well and look so good: Winnie Prentiss, Betsy Gammons, David Martinell, Cora Hawks, John Gettings, Alissa Cyphers, Karen Levy, and a cast of talented designers over the years. And then there are the beekeepers that have helped. Buzz Riopelle as an everready resource, Cindy Bee and Bill Owens who championed *Removing Bees* as a great way for beekeepers to add to their income, and Jennifer Berry who helped so much with the queen production section. Peter Sieling and Cindy Hodges added their artistic skills to the work, too. And again, to Kathy, who puts up with all this, time and time again. Thank you, all.